MANAGING PROJECT QUALITY

The books in the Project Management Essential Library series provide project managers with new skills and innovative approaches to the fundamentals of effectively managing projects.

Additional titles in the series include:

Managing Projects for Value, John C. Goodpasture

Effective Work Breakdown Structures, Gregory T. Haugan

Project Planning and Scheduling, Gregory T. Haugan

Project Measurement, Steve Neuendorf

Project Estimating and Cost Management, Parviz F. Rad

Project Risk Management: A Proactive Approach, Paul S. Royer

MANAGEMENTCONCEPTS
www.managementconcepts.com

MANAGING PROJECT QUALITY

Timothy J. Kloppenborg

Joseph A. Petrick

MANAGEMENTCONCEPTS

Vienna, Virginia

ſſſ
MANAGEMENTCONCEPTS

8230 Leesburg Pike, Suite 800
Vienna, VA 22182
(703) 790-9595
Fax: (703) 790-1371
www.managementconcepts.com

Printed in the United States of America

Library of Congress Cataloging-in-Publication Data
Kloppenborg, Timothy J., 1953–
  Managing project quality/Timothy J. Kloppenborg, Joseph A. Petrick.
    p. cm.—(Project management essential library)
  Includes bibliographical references and index.
  ISBN 978-1-56726-141-7 (pbk)
    1. Project management. 2. Quality Assurance. I. Petrick, Joseph A., 1946-II.
Title. III. Series.

HD69.P75 K598 2002
658.4'04—dc21
                                    2001056273

About the Authors

Timothy J. Kloppenborg, Ph.D., PMP, is Associate Professor of Management at Williams College of Business, Xavier University, and President of Kloppenborg and Associates, a consulting and training company based in Cincinnati, Ohio, that specializes in project and quality management. He holds an MBA from Western Illinois University and a Ph.D. in Operations Management from the University of Cincinnati. He is a Certified Project Management Professional (PMP) and has been active in the Project Management Institute for more than 15 years. Dr. Kloppenborg has published in journals including *Project Management Journal, PM Network,* and *Quality Progress.* He is a retired quality assurance officer from the United States Air Force Reserve. Dr. Kloppenborg has served in many practitioner, research, and consulting capacities on construction, information systems, and research and development projects.

Joseph A. Petrick, Ph.D., SPHR, is Professor of Management at the Raj Soin College of Business, Wright State University, and CEO of Performance Leadership Associates and Organizational Ethics Associates, management training and organizational development consulting firms based in Cincinnati, Ohio. He holds an MBA in quality management/marketing from the University of Cincinnati and a Ph.D. from Pennsylvania State University. He is a Certified Senior Professional in Human Resources (SPHR) and was a Malcolm Baldrige National Quality Examiner. Dr. Petrick is co-author of *Total Quality in Managing Human Resources* (St. Lucie Press, 1995), *Total Quality and Organizational Development* (St. Lucie Press, 1997), and *Management Ethics: Integrity at Work* (Sage Publications, 1997). He has managerial and consulting experience in the private, public, and nonprofit sectors and has published articles in the project management and quality management fields.

Table of Contents

Preface

A need has existed throughout history for both project management and quality management. During the second half of the twentieth century, however, the level of professional attention to the two fields increased dramatically because of increasing competition and complexity. Both fields grew rapidly, but largely without explicit awareness and use of their joint resources. Two exceptions are: (1) a few quality practitioners and academics recognized that project management techniques could be used to plan and manage quality improvement projects, and (2) the Project Management Institute (PMI®), a professional group for project managers, recognized that quality is one of the essential knowledge areas for project managers.

This book explicitly links the two fields and reinforces their convergence. We believe that the quality context (organizational and environmental), processes, and tools are essential to project management success. In turn, project stages and activities are essential to implementing quality management success. It is equally important to manage quality processes within the project stages and to manage the project's impact on its external context. A successful project manager uses the activities and tools to increase quality within the project stages and helps shape the external organizational and environmental context so that it remains supportive of project success.

As a result of their temporary nature, managing projects is intrinsically different from managing ongoing operations. However, many quality concepts and techniques have been developed primarily for use in ongoing operations. In this book, we adapt many quality tools and concepts to meet the unique challenges of projects. The purpose of this book is to present a roadmap and tools for managing project quality.

This book is targeted at four primary audiences: practicing project professionals, practicing quality professionals, academic and consulting practitioners, and students interested in managing quality projects. The first intended audience for this book is practicing project professionals. We specifically address many of our suggestions to project managers, project sponsors, project core team members, project suppliers, and project customers. Each has several important roles to play in delivering quality projects.

The second intended audience for this book is practicing quality professionals. Many quality practitioners already know how to use classic approaches to manage quality in an ongoing operation. Since most of these people will also be involved in some project work, this book can be useful to help them adapt standard quality practices for use on projects.

The third intended audience for this book is academic and consulting professionals. Researchers, educators, trainers, project consultants, and organizational change agents can benefit from increased sophistication in managing project quality.

The fourth intended audience for this book is students interested in managing quality projects. Students or associates in a formal training program can benefit from the structured integration of project and quality management provided by this book.

For all of these audiences, this book is valuable at each of four levels of learning, as described in the Kirkpatrick model.[1] For those at the first learning level of *unconscious incompetence* (i.e., you don't know that you don't know), this book provides a structured introduction to best practices to create basic awareness of the value of both fields. For those at the second learning level of *conscious incompetence* (i.e., you realize that you do not know), this book offers specific assessments, activities, and tools to instill deeper awareness and provide preliminary skills. We think many professionals who know either quality or project management, but not both, may be at this level.

For those at the third learning level of *conscious competence* (i.e., you know and do, but only with conscious effort), this book provides assessments, activities, roadmaps, and tools to increase skill competence by integrating the two fields in a newly developed five-stage model. Finally, for those at the fourth learning level of *unconscious competence* (i.e., effortless mastery), this book can help you make the transition from being an expert performer to being a skilled mentor who can explicitly share his or her competency with others to build a learning organization.

We hope that this book will help experts sustain learning organizations, deepen professional association learning, and expand domestic and global social learning about managing project quality.

---

[1]D.L. Kirkpatrick, *A Practical Guide for Supervisory Training and Development* (Reading, MA: Addison-Wesley, 1971).

In Chapter 1 of this book, we first briefly review both the project management and quality management fields. We next develop a detailed understanding of the four pillars of project quality management: customer satisfaction, process improvement, fact-based management, and empowered performance. Finally, we delineate the need for improvement in managing project quality.

The next five chapters of the book each represents one stage in the newly developed five-stage project quality management model: project quality initiation, project quality planning, project quality assurance, project quality control, and project quality closure. Each stage has a defined starting and ending point, with a sequence of activities and appropriate tools that would normally be used to manage project quality successfully.

The activities we describe are at a level of detail for a "middle of the road" project. A project that is simple, short, and familiar could streamline the manner in which the activities are completed, but would still need to accomplish the spirit of them. A large, complex, or unfamiliar project would need to perform the activities we describe, but in more detail. We feel this "middle of the road" approach will give project participants a good starting point from which to scale up or down.

Features included in this book to assist the reader include:

- An overall project flowchart to illustrate the five-stage project quality management model
- A detailed flowchart that shows the flow of activities within each stage
- A table at the start of each chapter that shows the four project quality pillars, activities, and tools
- Italicized concepts in text to visually highlight key ideas
- Chapter section numbers that correspond with the activities listed in each table
- Figures to help the reader visualize appropriate concepts and tools
- An integrated project quality activity matrix to summarize and highlight the core activities that require extra attention.

Timothy J. Kloppenborg
Joseph A. Petrick

Acknowledgments

W e would like to thank Cathy Kreyche, New Product Development Editor, and Myra Strauss, Managing Editor, at Management Concepts, for their helpful comments and support. We also thank our respective departments and universities (Management and Entrepreneurship Department, Williams College of Business at Xavier University, and Management Department, Raj Soin College of Business at Wright State University) for the support and encouragement they provided.

I (Tim) thank the project management mentors I have had: Rick Guenther and Denny Evans on project management practice and Sam Mantel and Dave Cleland on project management research.

Finally we thank our wives, Elizabeth Kloppenborg and Kimberly Petrick, and our children, Kathryn and Nicholas Kloppenborg, for their patience, understanding, love, and support, which made this book possible.

The extent to which *Managing Project Quality* succeeds in providing structured guidance and useful tools to our audiences is our ultimate measure of success. Please let us know both how this book has helped you in your work or studies and where you think it could be improved. We appreciate and welcome all your comments.

Timothy J. Kloppenborg
kloppenb@xu.edu
513-745-4905 (office)
513-745-4383 (fax)
www.xu.edu/management_dept/faculty/kloppenborg.htm

Joseph A. Petrick
joseph.petrick@wright.edu
937-775-2428 (office)
937-775-3545 (fax)
www.wright.edu/~joseph.petrick

CHAPTER 1

Introduction to Project Quality Management

P roject quality management is the combination of two fields: quality management and project management. Many factors—such as external global competitiveness, dynamic environmental changes, increased task complexity, and internal productivity improvement—have driven the parallel and separate evolution of quality management and project management. Superior quality and project management optimize the performance excellence of organizations, but their combined leverage is often underutilized. Quality processes can be used to improve project performance. Leaders who master project quality management will have greater success both on individual projects and on a portfolio of projects for their organizations.

An introduction to project quality management requires a basic understanding of: (1) the histories of the quality management and project management fields; (2) the conceptual foundations of project quality management; and (3) the need for improvement in project quality management.

BRIEF HISTORIES OF QUALITY AND PROJECT FIELDS

The histories of quality management and project management provide a context for understanding their interrelationships.

History of Quality Management

Before the Industrial Revolution, skilled craftspeople made and inspected their own limited number of products and took pride in their holistic workmanship before selling to their customers. After the Industrial Revolution, unskilled workers were employed in an assembly-line manufacturing system that valued quantity of output, specialization of labor, and separation of worker from customer. Nevertheless, concern for efficient quality control persisted because military and civilian customers objected to substandard product variations, such as weapons that did not function in combat and telephones that did not function in the home.

To address civilian concerns about variation in telephone service in the 1920s, Walter Shewhart's team at Bell Telephone Laboratories developed new theories and statistical methods for assessing, improving, and maintaining quality. Control charts, acceptance sampling techniques, and economic analysis tools laid the foundation for modern quality assurance activity and influenced the work of W. Edwards Deming and Joseph M. Juran.

After World War II, Deming and Juran introduced *statistical quality control* to the Japanese as part of General MacArthur's industrial base rebuilding program. They convinced top Japanese leaders that continually improving product quality through reducing statistically measured variation would open new world markets and ensure Japan's national future. From the 1950s to the 1970s, the Japanese improved the quality of their products at an unprecedented rate while Western quality standards remained stagnant. The Japanese were culturally assisted by the Deming Prize, which was instituted in 1951 by the Union of Japanese Scientists and Engineers (JUSE) to nationally recognize individuals and organizations that documented performance improvements through the application of company-wide quality control (CWQC). Starting in the late 1970s, the Japanese captured significant global market shares of the automobile, machine tool, electronics, steel, photography, and computer industries, in large part due to the application of quality management processes.

In a belated response to this quality-based, competitive threat from Japan, many U.S. organizations engaged in extensive quality improvement programs in the 1980s. In 1987—some 34 years after Japan created the Deming Prize—Congress established the Malcolm Baldrige National Quality Award (MBNQA), which provided a framework of seven categories (leadership, strategic planning, customer and market focus, information and analysis, human resource focus, process management, and business results) to promote quality management practices that lead to customer satisfaction and business results. In 1987 as well, the International Organization for Standardization (ISO) adopted written quality system standards (the ISO 9000 family of standards) for European countries and those seeking to do business with those countries, and later enacted a registration procedure. These design, development, production, installation, and service standards have been adopted in the United States by the American National Standards Institute (ANSI) with the endorsement and cooperation of the American Society for Quality (ASQ). In 1991, the European Foundation for Quality Management (EFQM), in partnership with the European Commission and the European Organization for Quality, announced the creation of the European Quality Award to signal the importance of quality in global competition and regional productivity.

The integration of these quality approaches at all organizational levels was referred to as Total Quality Management (TQM) in the 1990s and continues today, along with a recent emphasis on bottom-line, focused Six Sigma quality—a level of quality representing no more than 3.4 defects per million process opportunities.

History of Project Management

At the same time that quality management was developing, many events led to the need for better project management. While projects have occurred throughout history (for example, Egyptian pyramid construction projects, Chinese garden design projects, Roman road construction projects), the need for a systematic field of study emerged in the middle of the twentieth century in the United States. In the 1950s and 1960s, task complexity in dynamic environments in the defense, aerospace, construction, high-technology engineering, computer, and electronic instrumentation industries demanded formal project management skills at many levels. Previously, project management had been ad hoc at best. Now the need to address cost, schedule, scope, and quality concerns simultaneously forced companies and government organizations to develop more systematic and standard approaches.

In 1969, the Project Management Institute was formed to act as a forum for the discussion and exchange of project management experiences in different industries. In the 1970s and 1980s, the wide range of factors that prompted formal project management techniques surfaced: size of the undertaking beyond traditional functional resources, unfamiliarity of diverse efforts (e.g., crisis situations, takeover threats, major reorganizations), rapid market changes that put a premium on flexible, timely responsiveness, the interdependence and resource sharing necessary for the simultaneous engineering of new product innovations, and ad hoc team cooperation necessary to capitalize on a unique opportunity in conditions of uncertainty.

In 1981, the Project Management Institute formally recognized the development of uniform standards for management of projects as its responsibility and in 1987 it published *A Guide to the Project Management Body of Knowledge (PMBOK® Guide)*. Throughout all updated versions of the *PBMOK® Guide*, project quality management has been recognized as a separate, core knowledge area. Individuals who master the *PMBOK® Guide* and pass certification testing become Certified Project Management Professionals (PMP®).

Other trends in the 1980s and 1990s increased support for project management skills. For example, project management teams were used to implement quality management process improvements, concurrent engineering required better scheduling techniques, decentralized change management and risk management decisions in restructured firms highlighted the contribution of the field project manager as opposed to the traditional middle manager, and the distinctive needs of co-located and multinational teams on ad hoc assignments favored project management structures. In addition, the expansion of project-driven techniques from divisions such as management information systems (MIS) and research and development (R&D) to marketing and engineering has pressured many organizations to shift from traditional, long-lived product management structures to more flexible, short-lived project management structures.

CONCEPTUAL FOUNDATIONS OF PROJECT QUALITY MANAGEMENT

To understand these modern approaches in managing project quality, one must first understand the conceptual foundations of both quality management and project management. We cover those foundations next, followed by the four major project quality pillars that emerge from the conceptual foundations: (1) customer satisfaction; (2) process improvement; (3) fact-based management; and (4) empowered performance.

Conceptual Domain of Quality Management

One of the earliest approaches to project quality management occurred in ancient Babylon. During Hammurabi's rule, if a building collapsed, the architect and builder were both put to death. Fortunately, in modern times we focus more on preventing problems than claiming retribution if problems occur.

Quality has been defined as "the totality of characteristics of an entity that bear on its ability to satisfy stated or implied needs."[1] The stated and implied quality needs are inputs into devising project requirements. However, quality and grade are not the same. According to the *PMBOK® Guide*, grade is "a category or rank given to entities having the same functional use but different technical characteristics."[2]

Quality is a focus of project management. For example, a multimedia software program may be of high quality (no operational dysfunctions and an accurate accompanying manual) but be a low grade (a limited number of extra features). The mix of quality and grade is a responsibility of the project manager and his/her team.

Customer quality expectations in the manufacturing sector typically include the following factors:[3]

- Performance – A product's primary operating characteristics
- Features – The "bells and whistles" of a product
- Reliability – The probability of a product surviving over a specified period of time under stated conditions of use
- Conformance – The degree to which physical and performance characteristics of a product match pre-established standards
- Durability – The amount of use one gets from a product before it physically deteriorates or until replacement is preferable
- Serviceability – The ability to repair a product quickly and easily
- Aesthetics – How a product looks, feels, sounds, tastes, or smells
- Perceived quality – Subjective assessment resulting from image, advertising, or brand names.

Customer quality expectations in the service sector typically include the following factors:

- Time – How much time must a customer wait?
- Timeliness – Will a service be performed when promised?
- Completeness – Are all items in the order included?
- Courtesy – Do front-line employees greet each customer cheerfully and politely?
- Consistency – Are services delivered in the same fashion for every customer, and every time for the same customer?
- Accessibility and convenience – Is the service easy to obtain?
- Accuracy – Is the service performed right the first time?
- Responsiveness – Can service personnel react quickly and resolve unexpected problems?

Since meeting or exceeding customer expectations and conforming to system design and specifications are crucial to quality, the analytical framework offered by the quality performance grid (see Figure 1-1) is helpful in depicting the relative parameters of achieved quality. In the grid, the vertical axis represents managerial performance quality with respect to meeting customer satisfaction expectations. The horizontal axis represents technical performance quality with respect to meeting design and system specifications. World-class quality requires high level (90 percent) mastery of both managerial and technical skills. Less than 50 percent success in either

FIGURE 1-1 Quality Performance Grid

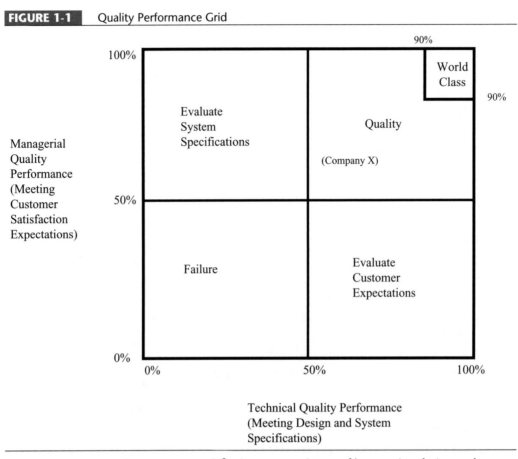

meeting customer satisfaction expectations and/or meeting design and system specifications is considered a quality performance failure.[4]

Company X is shown as an example. Company X has satisfactory performance in both dimensions, but is far from world class. This quality performance grid can be used to ensure that a company is performing satisfactorily on both the managerial and technical dimensions of quality. It can also be used to identify where more effort is needed.

The *cost of poor quality* is the total amount of money a company spends to prevent poor quality (i.e., to ensure and evaluate that the quality requirements are met) plus any other costs incurred as a result of poor quality being produced.[5] Poor quality can be defined as waste, errors, or failure to meet customer needs and system requirements.

The costs of poor quality can be broken down into the three categories of prevention, appraisal, and failure costs.

- **Prevention costs:** These are planned costs an organization incurs to ensure that errors are not made at any stage during the delivery process of that product or service to a customer. The delivery process may include design, development, production, and shipping. Examples of prevention costs include quality planning costs, information systems costs, education and training costs, quality administration staff costs, process control costs, market research costs, field testing costs, and preventive maintenance costs. The costs of preventing mistakes are always much less than the costs of inspection and correction.
- **Appraisal costs:** These include the costs of verifying, checking, or evaluating a product or service during the delivery process. Examples of appraisal costs include receiving or incoming inspection costs, internal production audit costs, test and inspection costs, instrument maintenance costs, process measurement and control costs, supplier evaluation costs, and audit report costs.
- **Failure costs:** A company incurs these costs because the product or service did not meet the requirements and had to be fixed or replaced, or the service had to be repeated. These failure costs can be further subdivided into two groups: internal or external failures.

    Internal failures include all costs resulting from the failures found before the product or service reaches the customer. Examples include scrap and rework costs, downgrading costs, repair costs, and corrective action costs from nonconforming product or service.

    External failures occur when the customer finds the failure. External failure costs do not include any of the customer's personal costs. Examples of these failure costs include warranty claim costs, customer complaint costs, product liability costs, recall costs, shipping costs, and customer follow-up costs.

Conceptual Domain of Project Management

Understanding the concepts of quality management is important as a basis for learning project quality management. Now we look briefly at the basics of project management. *Projects* are defined in the *PMBOK® Guide* as "temporary endeavors undertaken to create a unique product or service."[6] The objectives of projects and operations are fundamentally different from a timing perspective. The focus of the project is to quickly achieve the objective and then terminate. The objective of an ongoing non-project operation is to sustain itself and the organization indefinitely.

A successful project is one that meets at least four criteria: schedule, budget, performance, and customer satisfaction. In other words, successful projects are those that come in on time, on budget, perform as expected by conforming to design specifications, and satisfy customers.

Since the 1980s and 1990s, project managers and their teams have been used to implementing quality management process improvements by relying on project lifecycles. While there are a variety of generic project lifecycle models, the authors have developed a new *five-stage project quality process model*, presented in Figure 1-2. The first and last stages are not currently in the *PMBOK® Guide,* but are crucial to project quality success and parallel other *PMBOK® Guide* recommendations for other core knowledge areas.[7]

The five stages are:
1. Project quality initiation
2. Project quality planning
3. Project quality assurance
4. Project quality control
5. Project quality closure.

We believe this five-stage model is the simplest generic model that can be used to show when, why, and how critical quality management techniques can be effectively used to help ensure project success. We believe that all five stages are needed, even though some managers frequently shortchange one or two of them. We also believe that this model can be used for projects in any industry. Additional or more detailed quality management techniques may be needed in some industries and on large, complicated projects in any industry. We believe that managers of even the smallest, simplest projects should understand the need for all five stages and the quality management techniques we suggest for each. If a manager wants to use a streamlined approach on a simple project, that is fine—as long as he or she accomplishes the spirit of the techniques shown.

In our five-stage project quality process model, we show the relationships between each stage. For simplicity, we are showing only the starting and ending points of each stage. In the following chapters we will show and discuss the many activities that should occur during each stage.

The first stage, *project quality initiation*, begins with the identification of a potential project and ends with a signed authorization to proceed. The second stage, *project quality planning*, begins with the signed authorization to proceed and ends with the acceptance of the project plan by stakeholders. The third stage, *project quality assurance*, begins with the acceptance of the project plan by stakeholders and ends with processes and deliverables improved to

---

FIGURE 1-2   Five-Stage Project Quality Process Model

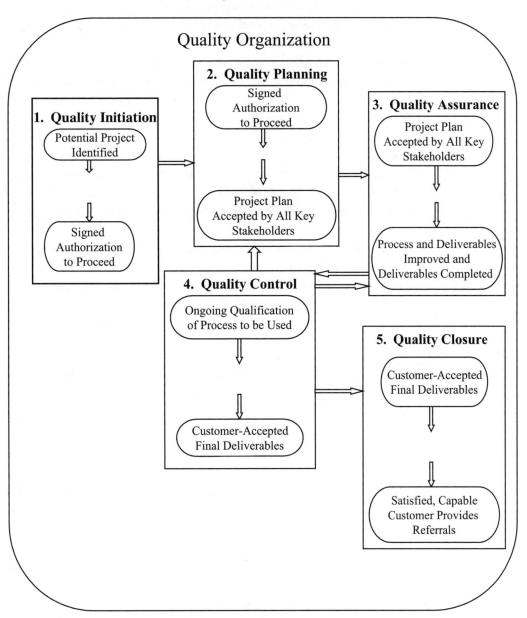

Quality Environment

Quality Organization

2. Quality Planning

Signed Authorization to Proceed

1. Quality Initiation

Potential Project Identified

Signed Authorization to Proceed

Project Plan Accepted by All Key Stakeholders

3. Quality Assurance

Project Plan Accepted by All Key Stakeholders

Process and Deliverables Improved and Deliverables Completed

4. Quality Control

Ongoing Qualification of Process to be Used

Customer-Accepted Final Deliverables

5. Quality Closure

Customer-Accepted Final Deliverables

Satisfied, Capable Customer Provides Referrals

Quality Context = Quality Organization + Quality Environment

the point of completion. The fourth stage, *project quality control*, begins with the ongoing qualification of processes used and ends with client acceptance of the final deliverables. Hence, the technical quality performance of meeting project design specifications occurs primarily in the third stage (project quality assurance) and the managerial quality performance of satisfying project customers occurs primarily in the fourth stage (project quality control). The third and fourth stages are not sequential as are stages one, two, and five; they are dynamically interactive and interdependent. The fifth stage, *project quality closure*, begins with the client acceptance of the final deliverables and ends with referrals from a capable, satisfied customer.

Leaders of organizations need to determine who will perform each project task. We show a typical set of project quality management role assignments in the project lifecycle accountability matrix presented in Figure 1-3. While a leader will use many factors to determine who performs each task, this accountability matrix can serve as a useful starting point in making role assignments.

Now that we have considered the basic concepts of quality and project management separately, we put them together. We feel the best way to understand the combined field of project quality management is to describe it as the sum of four pillars: (1) customer satisfaction, (2) process improvement, (3) fact-based management, and (4) empowered performance.[8] Each pillar must be strong to hold up the project as a pillar holds up a building.

FIRST PROJECT QUALITY PILLAR: CUSTOMER SATISFACTION

The first project quality pillar is the strategic priority accorded customer satisfaction, which is achieved by customer-focused work systems supported by committed leadership. Meeting both external and internal customer expectations drives strategic efforts in a quality firm.

For purposes of clarification, a number of conceptual distinctions must be made at the outset. The first clarification is between project stakeholders and project customers. *Project stakeholders* can be defined as those directly or indirectly associated with the project, those affected in the long/short term by the project and its activities, and those interested in the outcome of the project. Often project stakeholders are divided into internal and external stakeholders. Internal stakeholders typically include members of the home organization: the project sponsor, the project manager, the project team, top management, functional managers, staff personnel, service and support, other project managers, and internal subcontractors. External stakeholders typically include: customers/clients, suppliers, distributors, regulatory agen-

FIGURE 1-3   Project Lifecycle Accountability Matrix

Role\Stage	Project Quality Initiation	Project Quality Planning	Project Quality Assurance	Project Quality Control	Project Quality Closure
Sponsor	Select project manager, align and select project, commit to charter	Determine decision-making authority, commit to plan	Conduct external customer communications, mentor project manager, and clear obstacles as needed	Conduct external customer communications, mentor project manager, and clear obstacles as needed	Recognize and reward participants, assess project to improve system
External Customer	Identify and prioritize expectations, commit to charter	Identify customer satisfaction standards and tradeoff values, commit to plan	Conduct ongoing communications	Confirm ongoing satisfaction level, accept deliverables	Verify when training and support are complete, assess project to improve system
Project Manager	Select core team, identify risks, empower performance, commit to charter	Identify customer satisfaction standards and tradeoff values, develop quality and communications plans, commit to plan	Conduct external customer communications, confirm qualified processes used, manage quality audits and planning	Measure customer satisfaction, manage process improvements	Recognize and reward participants, assess project to improve system
Core Team	Determine team operating principles, flowchart project, identify lessons learned, commit to charter	Plan project, identify suppliers, qualify the process, identify data to collect, commit to plan	Use qualified processes, gather data, find root causes, conduct quality audits, plan future work	Measure customer satisfaction, test deliverables, correct defects, endorse deliverables	Provide customer support and training, assess project to improve system

cies, social and cultural environment, economic and financial environment, political and legal environment, external contractors and competitors, media and public interest groups, and the natural ecological environment.

Project customers are the direct purchasers, end users, and providers of products and services. Project customers are also both internal and external. The external customer is usually accorded highest priority in quality organizations; nevertheless, internal home organization customers must also be satisfied.

We will adopt the conventional phrase *key project stakeholders* to refer to that mix of internal and external direct purchasers, consumers, and providers referred to as *customers*. It is, therefore, customers or key project stakeholders who must be satisfied for the first project quality pillar to be established. It is advisable to satisfy as many additional stakeholders as possible to prevent any unwarranted project disruption.

Distinctions about the nature of satisfaction also need to be addressed. Distinctions have been made among product characteristics as being *dis-satisfiers, satisfiers,* and *exciters/delighters.* Dissatisfiers are unstated customer expectations for the product or service that are taken for granted and, if absent, result in customer dissatisfaction with products. Satisfiers are stated customer expectations about the product or service, which, if fulfilled, lead to product satisfaction. Exciters/delighters are unstated and unexpected consumer desires for products or services which, if met, lead to high perceptions of quality and likely purchase or repurchase of products.

Over time, exciters/delighters become satisfiers as customers become used to them, and eventually satisfiers become dissatisfiers. This means that systemic strategic planning and leadership are required to ensure that ongoing customer satisfaction is delivered as customer expectations increase.

A *work system* can be defined as a set of functions or activities within an organization that interact to achieve organizational goals. To engage in systemic strategic planning requires that leaders understand the interrelationships among all subsystem parts and the people who work in them. Deming specifically emphasizes that the leader's primary responsibility is to optimize the quality system so that customer satisfaction will result. By supporting projects that are best for one manager's career or for a highly vocal group, the leader suboptimizes. *Suboptimization* results in a net loss for the organization by diverting resources from system-aligned projects to marginal projects.

For example, a project manager and his/her team in the purchasing department may recommend the purchase of new materials at the lowest bid to cut costs. Inexpensive materials may be inferior in quality. This might cause excessive costs in later corrections during manufacturing. Although the purchasing project leader and team may look good on paper, the entire system will suffer. Therefore, an important responsibility of the committed quality leader is to ensure that only system-aligned projects are sponsored and completed in order to prevent suboptimization.

Quality strategic planning is the organizational design and structure that produces total customer satisfaction. Strategic planning results in both customer satisfaction *goals* (non-quantified aspirations) and customer satisfaction *objectives* (which determine what is to be accomplished by when in quantified terms).

Now that we understand who the various project customers are, what delights and satisfies them, and how to use strategic planning to best satisfy our mix of customers, we turn to our next project quality pillar.

SECOND PROJECT QUALITY PILLAR: PROCESS IMPROVEMENT

The second project quality pillar is the continual (includes both continuous and discontinuous) improvement of work processes to efficiently and effectively achieve the strategic goal of customer satisfaction. A *work process* can be defined as any set of linked activities that takes an input, adds value to it, and provides an output to an internal or external customer. Thus, a set of processes may together form a quality system. The quality system in turn provides the organizational operational context for team projects and individual task performances.

Ongoing process improvement results in three types of quality improvement: incremental cost reduction, competitive parity, and breakthrough dominance. All three types of improvement are important and each is appropriate in certain circumstances. Any given project is likely to use one or more of these types of improvement.

The first type, *incremental cost reduction* (sometimes referred to as *kaizen*), is the process improvement approach that constantly and gradually cuts costs and involves every organizational member in order to maintain the existing system more efficiently. An example is to reduce the number of steps in a process without sacrificing quality.

The second type, *competitive parity*, is the process improvement approach that abruptly and dramatically matches the performance of the best-in-class of external competitors. Strategic planners and key process champions usually drive this type of improvement; it may entail scrapping the existing system and rebuilding to catch up with the best-of-class. An example is Microsoft rebuilding its processes to match Internet competitors.

The third type, *breakthrough dominance*, is the process improvement approach that involves quantum leaps to outdistance the competition and revolutionarily restructure or reengineer new processes. Usually, strategic leadership, R&D management, and process change champions drive this type of improvement. It may entail starting over and creating a new system from scratch. An example is the radical redesign of jet engines to surpass propeller-driven aircraft.

Furthermore, process improvement entails *process qualification* determinations, as indicated in Figure 1-4. The goal is to move from:

1. The spontaneous level in which little or no process standards are used; through

FIGURE 1-4 Process Qualification Levels

Level 1–Spontaneous: Few or no process standards are used.
- Lack of documentation
- Skills and knowledge uneven
- Inadequate tracking
- Very little use of systems or technical tools
- Process success depends on experience and skills of managers and team

Level 2–Initialized: Process awareness is widespread but ad hoc.
- Non-standard methods and approaches widely used, everyone performs differently
- Some documented procedures (what needs to be done but not how to do it)
- Some data collection and documentation
- Technical tools used but not always in a full or correct manner
- All processes attempt to follow some basic functionality

Level 3–Formalized: Basic processes are standardized and institutionalized.
- Company-wide standards developed and documented for all basic processes to maintain an existing system
- Audited and enforced use of standard processes
- Consistent data collection and reporting across organization
- Lessons learned are shared throughout organization
- Widespread and adequate process specific training to keep current system functioning

Level 4–Optimized: Processes are systematically measured, continually improved, and cross-functionally integrated with business operations.
- Data consistently collected and stored in a database, and extensive evaluation performed for all processes
- Database integrated with company systems to ensure ongoing improvement
- Mechanisms established for continuous process improvement
- Innovative ideas pursued and organized to improve processes and documentation
- Goes beyond process success, emphasizes success of people and systems

2. The initialized level in which non-standard approaches are widely used; also through
3. The formalized level in which standards are institutionalized; and finally to

4. The optimized level in which improvement and integration are a way of life.

The four levels of process qualification provide both a multi-level classification scheme for existing processes and a "to-do" list for fact-based project management teams.

THIRD PROJECT QUALITY PILLAR: FACT-BASED MANAGEMENT

The third project quality pillar focuses on the importance of managing by facts rather than managing by power, hunches, or groupthink. To *manage by facts* means that an organization (1) uses quality processes to identify and capture data and trends that determine what is factually true about performance, and (2) structures itself to be responsive to diverse stakeholders that voice the truth. Collecting, measuring, and analyzing data and trends are key responsibilities for project leaders and teams in evaluating and improving processes.

One of the most important skills in fact-based management is knowledge of *statistical variation* in evaluating processes. Processes that include materials, tools, machines, operators, and the environment exhibit complex interactions; properly understanding them requires knowledge of two types of statistical variation.

One is *common or random variation*, which is inherent in any process. Multiple small causes are responsible for random variation. A system governed only by common causes is said to be *stable*. To decrease this type of variation one needs to improve the entire system, not just one part.

The second type of statistical variation is *special or assignable variation*. Assignable causes of variation occur when something in the process is different from normal, such as faulty material, an inattentive worker, or a broken tool. The way to reduce assignable causes of variation is to identify and control them as quickly as possible.

Statistical quality control charts (such as in Figure 1-5) are line graphs with center lines and statistically calculated upper and lower control limits used to distinguish between random and assignable cause variation. Work performance differences within the upper and lower control limits are statistically insignificant although they may appear to be important to those not skilled in fact-based management.

Project leaders can make two fundamental mistakes in attempting to improve a process without factual knowledge of its statistical variation. The first mistake is *overcontrol*—treating as a special cause any fault, complaint,

FIGURE 1-5   Project Control Chart

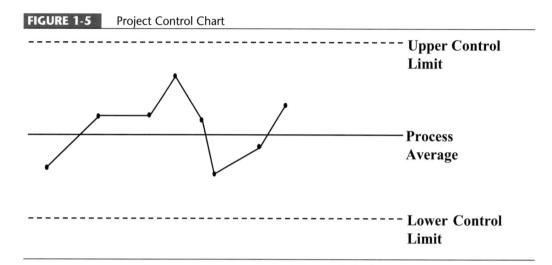

mistake, breakdown, accident, or shortage when it actually came from common cause. Some people call this the Dilbert effect of abusive managerial overcontrol. The second mistake is *undercontrol*—attributing to common causes any fault, complaint, mistake, breakdown, accident, or shortage when it actually comes from a special cause. Examples of undercontrol include neglecting to identify, retrain, or dismiss substandard performers at work.

In the case of overcontrol, interfering with a stable system actually increases variation and harms the system. In the case of undercontrol, project leaders miss the opportunity to eliminate unwanted variation by assuming that it is uncontrollable. Since producers and consumers benefit from reduced variation, project managers and team members need knowledge of statistical variation to properly manage by facts.

Another important group of skills in fact-based management is those necessary to lead and follow in a variety of teams, including cross-functional teams. These skills contrast sharply with those needed merely to respond to hierarchical authority. Leading and following skills are crucial for the decentralized and horizontal management of information streams in the organization. *High-quality project teams* move from initial project awareness, to involvement, to commitment, and finally to project ownership on their own or with the skilled intervention of seasoned project managers. They are successful and rapidly socialize new members into performance norms of cooperative competence and power-sharing.

The third project quality pillar of fact-based management leads right into the fourth pillar: empowered performance.

FOURTH PROJECT QUALITY PILLAR: EMPOWERED PERFORMANCE

The fourth project quality pillar entails the empowered daily work performance of continual improvement in personal tasks aligned with the system and within an employee's scope of responsibility.

Work performance can be defined as behavior associated with the accomplishment of expected, specified, or formal role requirements on the part of individual organization members. Quality organizations may be described in terms of the norms, values, and reward procedures that emphasize the holistic, competent behavior of individuals oriented toward cooperation with fellow organization members.

Work performance in a quality environment includes accomplishing tasks and taking initiatives above and beyond the call of duty, along with sharing information with and helping co-workers. This performance is typically referred to as *organizational citizenship behavior* (OCB). In a total quality organization, OCB is both expected and formally rewarded. Support staff in a quality office will often phone other departments for work if their own department's work has been completed. This cooperative "helping out" attitude is the recognized norm and is routinely celebrated and rewarded.

Individual empowerment has been described as intrinsic task motivation consisting of five dimensions: responsible choice, meaningfulness, competence, proactive learning, and impact. The central component of empowerment is responsible choice—free decisions for which one is responsible. Choice involves taking responsibility for a person's actions. Choice also develops an individual's belief in his/her ability to effect a desired change in the environment. Field research has demonstrated that choice and personal control are related to intrinsic task motivation, job performance, and job satisfaction.

The second dimension, meaningfulness, concerns the value a task holds for the individual. If an individual finds a task meaningful, he or she will be more content performing it. The third dimension, competence, refers to the experientially founded belief that one is capable of successfully performing a particular task or activity. People who believe they can perform the work assigned are more willing workers.

The fourth dimension, proactive learning, is the process of discovering, creating, and/or understanding through feedback between practices and results. Empowered people are used to and expect feedback. They are not overly sensitive to critical remarks. The fifth dimension, impact, represents the degree to which individuals perceive that their behavior makes a difference.

Project leaders should think about all five dimensions of individual empowerment as they deal with project participants. Often short conversations regarding one or more of these dimensions can help individuals feel more empowered, thereby improving the chances for good quality work on the project.

Individuals usually appreciate organizations that provide them with opportunities for personal control, responsibility, and challenge in their work, and will tend to reciprocate by being more committed to the organization. As individuals demonstrate empowerment readiness in project responsibilities, they develop their sense of self-respect through performance.

Quality firms require respect for all people in the organization, regardless of role, since each person is continually being empowered to enhance the effectiveness of the organization. We now describe several problems that deal with lack of respect.

Individuals who respect others but not themselves are a problem. Unfortunately, these individuals do not relate well to others in a cooperative quality manner because they undervalue their own worth, rarely voice their own opinions, and rely on the approval of others for validation. An example is a project leader or team member who allows others to verbally abuse him/her without setting boundaries for respectful discussion at work.

Another problem concerns individuals who respect themselves but not others. They alienate team members and are unable to learn from others or to generate teamwork. An example is project leaders who do not solicit input or ignore feedback from knowledgeable followers because they (the leaders) are too proud to learn from others.

Yet another lack of respect problem is that some people only feel or show honor for those who have higher rank or status in work organizations and treat peers or direct reports with contempt or neglect. Some people profess respect for others, but act as if they always expect others to defer to their judgment. For example, they often dismiss the contributions of others in conversations and decision-making processes. This gap between the rhetoric and reality of respect for people is what must be—and is—eliminated or severely reduced in a quality organization because the system cannot improve without sincere respect for the integrity of individual contributions.

NEED FOR IMPROVED PROJECT QUALITY MANAGEMENT

Now that we have discussed the conceptual domains of the quality and project management fields and the four project quality pillars, it is easy to see why lack of familiarity with both fields can cause problems. Failure to understand and use both project and quality tools may lead to many problems. First we consider potential problems that may arise if people do not understand the four project quality pillars in general and then we consider potential problems that may arise if people do not understand the activities that are required during each of the five stages of project quality management.

When people do not understand and/or use the first project quality pillar, customer satisfaction, they:
- Do not strategically prioritize customer satisfaction and instead often prioritize short-term financial returns and wonder why they are losing market share
- Do not understand systems so they see events as isolated incidents rather than the net result of many interactions and interdependent forces
- Confuse operational symptoms with deeper dysfunctional system causes
- Sponsor projects that suboptimize resources and thereby dissipate the energy of the firm.

When people do not understand and/or use the second project quality pillar, process improvement, they:
- Regard only the efficient maintenance of status quo operations, rather than additional ongoing process improvement, as the ideal work contribution
- Cannot distinguish between different levels of process qualification so they cannot optimize organizational performance.

When people do not understand and/or use the third project quality pillar, fact-based management, they:
- Overcontrol people who are performing acceptably in a stable system and thereby reduce system productivity and lower morale
- Undercontrol people who are statistically substandard performers and miss opportunities to rid the system of unwanted variation
- Are unable or unwilling to cooperatively engage in cross-functional teamwork to improve processes.

When people do not understand and/or use the fourth project quality pillar, empowered performance, they:

- Engage in workplace avocations that divert their energy into non-aligned activities that waste team and organizational resources
- Spend too much time trying to get individual recognition and never develop the teamwork skills to constructively contribute to collective projects for process improvement
- Do not develop individual empowerment skills and respectful regard for others' competencies so they resort to dominance or victimization rituals that are personally and organizationally counterproductive.

Now we turn our attention to some of the problems that may be encountered when people do not understand the different stages of project quality management. When people do not understand project quality initiation, they

- Endorse suboptimal projects that should not be initiated
- Poorly understand the potential project
- Generate insufficient support.

When people do not understand project quality planning, they:

- Ignore needed inputs and suppliers
- Neglect to qualify project processes
- Do not secure the necessary project commitments.

When people do not understand project quality assurance, they:

- Do not confirm that qualified processes are being used
- Do not gather sufficient data
- Do not improve work process execution
- Mismanage the human resource subsystem.

When people do not understand project quality control, they:

- Inadequately measure customer satisfaction
- Insufficiently test products against standards
- Inadequately perform statistical analyses of problem causes so that final deliverables do not meet customer expectations.

When people do not understand project quality closure, they:

- Do not provide for customer capability through training and support
- Fail to recognize and reward participants
- Neglect to collect and share lessons learned with other organization members.

To address these pressing needs for improving overall project quality management, we now show how the four project quality pillars can be applied during each of the five stages of project quality management. Each stage will be covered in one of the following chapters, starting with Chapter 2: Project Quality Initiation.

NOTES

1. International Organization for Standardization (ISO), *Quality Management and Quality Assurance* (Geneva, Switzerland: ISO Press, 1994).
2. Project Management Institute Standards Committee, *A Guide to the Project Management Body of Knowledge* (*PMBOK® Guide*) (Upper Darby, PA: Project Management Institute, 2000), p. 96.
3. James R. Evans and William M. Lindsay, *The Management and Control of Quality*, 5th edition (Cincinnati, OH: South-Western Publishing, 2002).
4. Jeffrey S. Leavitt and Philip C. Nunn, *Total Quality through Project Management* (New York: McGraw-Hill, 1994).
5. Philip B. Crosby, *Quality Is Free: The Art of Making Quality Certain* (New York: Dutton, 1979).
6. Project Management Institute Standards Committee, *A Guide to the Project Management Body of Knowledge* (*PMBOK® Guide*) (Upper Darby, PA: Project Management Institute, 2000), p. 4.
7. Ibid., p. 38.
8. Joseph A. Petrick and Diana S. Furr, *Total Quality in Managing Human Resources* (Delray Beach, FL: St. Lucie Press, 1995); William M. Lindsay and Joseph A. Petrick, *Total Quality and Organization Development* (Delray Beach, FL: St. Lucie Press, 1997).

CHAPTER 2

Project Quality Initiation

A project normally begins with a potential project being identified. For our purposes, it does not matter where the idea originated—just that there is a potential project. Project quality initiation, therefore, begins with the identification of a potential project and ends with a signed authorization to proceed. *Initiation* is defined in the *PMBOK® Guide* as "the process of formally recognizing that a new project exists or that an existing project should continue into its next phase."[1] Project quality initiation is the first stage of the five-stage project quality process model, as depicted in Figure 2-1.

The quality context of the model shows that both the organization and the environment can impact the project. While some projects involve the interface of the organization and the environment, the five-stage structure of project processes usually remains the same. Whether the project involves organizational change or organizational stability, Figure 2-2 identifies the flowchart of activities entailed in this stage.

As discussed in Chapter 1, an effective project participant needs an understanding of the four quality pillars and an ability to use various project quality tools to complete specific project quality activities. These pillars, activities, and tools facilitate the movement from the initial identification of a potential project to the signed authorization to proceed with a project. Table 2-1 categorizes the project quality pillars, activities, and tools for the quality initiation stage into a project factors table.

This chapter will follow the order of project quality pillars and their sequenced activities in Table 2-1: (1) customer satisfaction, (2) process improvement, (3) fact-based management, and (4) empowered performance. These are the same four pillars of project quality introduced in Chapter 1. The first number of the listed activities corresponds to the appropriate project quality pillar, e.g., Activity 1.1 is associated with Pillar 1 and Activity 2.1 is associated with Pillar 2. The second number refers to the typical approximate chronological sequence of its execution within the pillar's domain, although

FIGURE 2-1   Five-Stage Project Quality Process Model

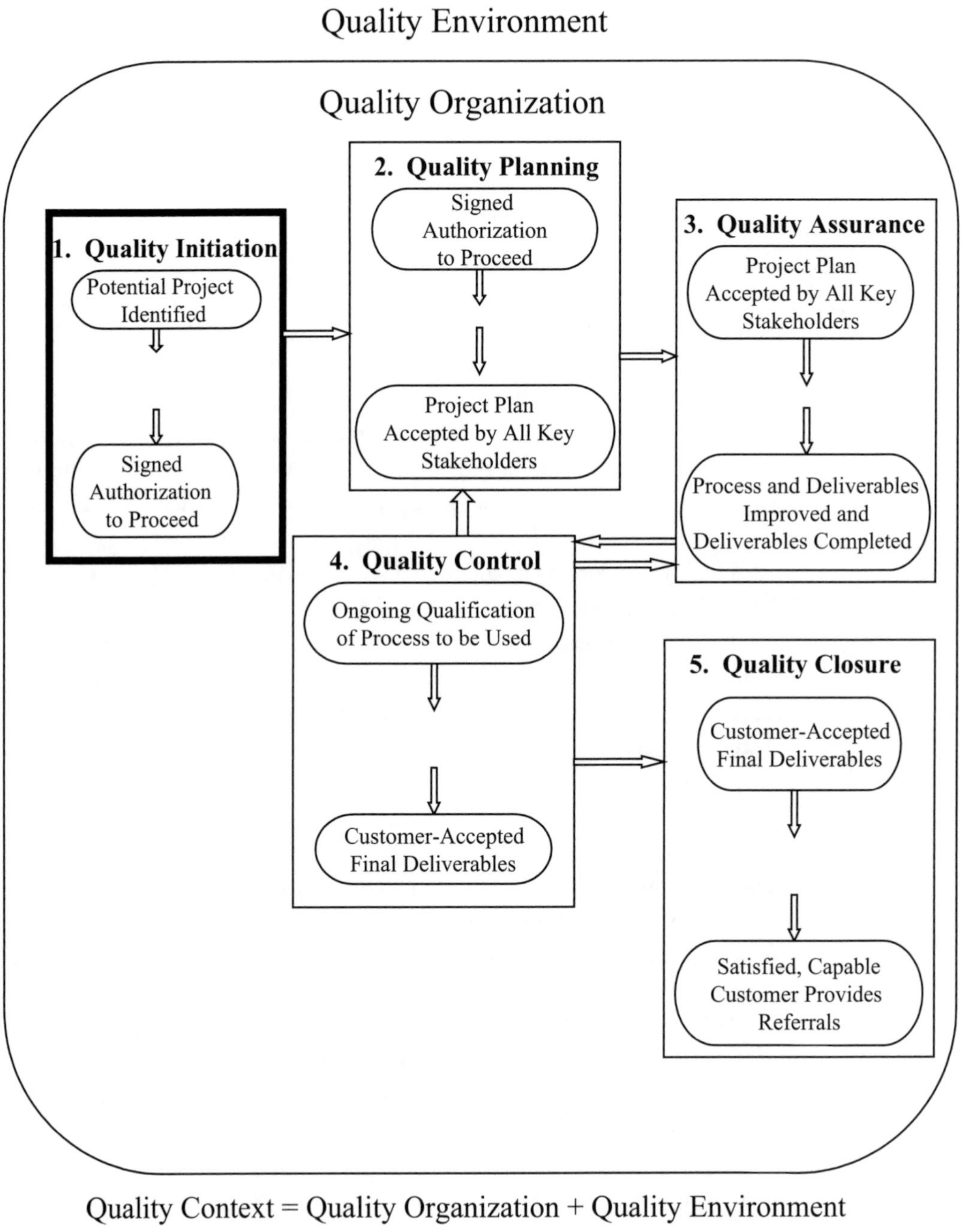

Quality Environment

Quality Organization

2. Quality Planning

Signed Authorization to Proceed

1. Quality Initiation

Potential Project Identified

Signed Authorization to Proceed

Project Plan Accepted by All Key Stakeholders

3. Quality Assurance

Project Plan Accepted by All Key Stakeholders

Process and Deliverables Improved and Deliverables Completed

4. Quality Control

Ongoing Qualification of Process to be Used

Customer-Accepted Final Deliverables

5. Quality Closure

Customer-Accepted Final Deliverables

Satisfied, Capable Customer Provides Referrals

Quality Context = Quality Organization + Quality Environment

FIGURE 2-2 Project Quality Initiation Flowchart

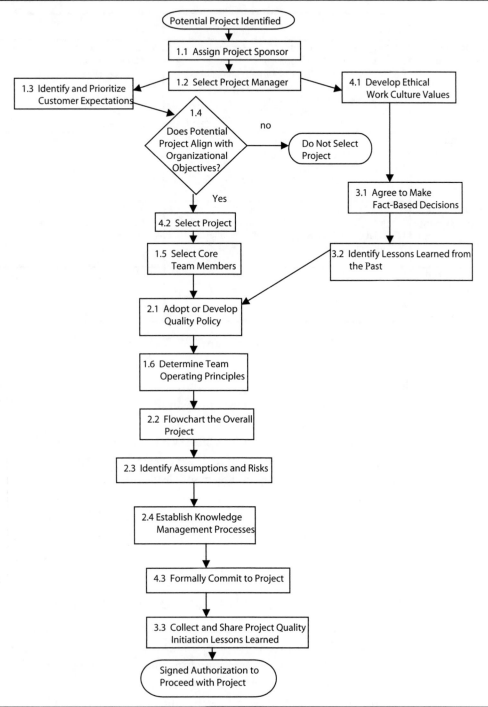

TABLE 2-1 Project Quality Initiation Factors Table

Pillars	Activities	Tools
1. Customer Satisfaction	1.1 Assign Project Sponsor 1.2 Select Project Manager 1.3 Identify and Prioritize Customer Expectations 1.4 Align Project with Organizational Objectives 1.5 Select Core Team Members 1.6 Determine Team Operating Principles	Participant Readiness Assessment Participant Readiness Assessment House of Quality Participant Readiness Assessment
2. Process Improvement	2.1 Adopt or Develop Quality Policy 2.2 Flowchart the Overall Project 2.3 Identify Assumptions and Risks 2.4 Establish Knowledge Management Processes	 Flow Chart PDCA Model
3. Fact-Based Management	3.1 Agree to Make Fact-based Decisions 3.2 Identify Lessons Learned from the Past 3.3 Collect and Share Project Quality Initiation Lessons Learned	 Modes of Knowledge Conversion Plus Delta Model
4. Empowered Performance	4.1 Develop Ethical Work Culture Values 4.2 Select Project 4.3 Formally Commit to Project	Ethical Work Culture Assessment Project Charter

this sequential order may well vary with different projects, organizations, or industries. For example, 1.1 Assign Sponsor, normally comes before 1.2, Select Project Manager.

We have aimed our descriptions of the various activities at "middle of the road" projects. A leader of a complex, large, or unfamiliar project may need to use more detailed techniques. Likewise, a leader of a short, simple, familiar project may be able to streamline the techniques. We feel that a skilled leader can use our in-between level as a starting point and scale up or down. We also believe that leaders of even the smallest projects should understand the need for each activity we list before they streamline or they are likely to miss some essential project quality management activities.

FIRST PROJECT QUALITY PILLAR: CUSTOMER SATISFACTION

Project quality initiation begins with the identification of internal customers or prospective project participants. This is often an iterative process. It starts with identifying a project sponsor and a project manager. These individuals should then work with the external customers to identify and prioritize expectations and to ensure that the prospective project is aligned with organizational objectives to avoid suboptimization. Once this process is complete, the sponsor and project manager will know enough to select the project's core team members. The core team, with the project manager's guidance, will then determine the team's operating principles.

1.1 Assign Project Sponsor

The *project sponsor*, usually assigned by top management, will mentor the project manager and champion the project, help the project manager secure resources, and help remove obstacles to project progress. The sponsor should be a primary stakeholder. The sponsor needs to sell the project to top management. Often the sponsor is the person who most wants the project to be performed and will be the catalyst for proposing the project and getting it selected.

One tool for determining the project quality readiness of project sponsors, managers, and other key individuals is the *Project Quality Participant Empowerment Readiness Assessment* (PERA) Instrument included as Appendix A. The PERA measures the perceived relative level of technical task maturity, administrative psychosocial maturity, and participant moral maturity that prospective sponsors, managers, and team members have. Using a 360-degree feedback process provides a broader set of judgments that is more likely to select and support project participants who will successfully complete projects.

1.2 Select Project Manager

The *project manager* will be operationally responsible for most of the project planning and execution. The project manager, selected after PERA feedback, is the operational driver of the project who is charged with the responsibility to complete the task. He or she uses a variety of management styles to ensure project success by addressing levels of technological uncertainty, system complexity, and professional ethics compliance. To reduce risks to project quality, it is advisable to select a competent, experienced project manager along with a competent, experienced project sponsor. Joining a rookie sponsor with a rookie project manager raises the risk of project quality problems. Both the selection and conduct of the project sponsor and project manager will set an example for the remainder of the project participants.

1.3 Identify and Prioritize Customer Expectations

Project customer satisfaction is the primary strategic focus of the quality initiation stage. Working toward this goal requires identifying and prioritizing customer expectations. Doing so in turn requires extensive collection and analysis of customer feedback data.

Start by learning the customer's working environment and the intended use of the project's output. This may involve visiting the customer. It is important to remember that different users within the customer organization may have different expectations. The customer's expectations (and, therefore, project requirements) will flow from the intended use to which the customer applies the project output.

Very early in the quality initiation stage, the project manager must understand and anticipate, at least at a high level, what the customer expects from the project. This is both to determine whether this potential project makes sense and should be selected, and to serve as the basis of more detailed understanding later.

One quality tool for identifying customer expectations and translating customer priorities into project specifications is the *project house of quality*, depicted in Figure 2-3.[2]

Building the house of quality for a project entails six basic steps:

1. Identify the customer's project desires/expectations
2. Identify project technical features/specifications
3. Determine the relative strength of relationships between the customer's expectations and project specifications, and interrelationships between project specifications
4. Conduct an evaluation of competing existing and potential projects
5. Obtain customer importance rankings to indicate relative priorities for determining key project selling points in comparison to competitive projects
6. Design in project specification priorities as voiced by the customer.

The project house of quality can also be used for several essential checks that the project team should perform on the customer's expectations. These checks include completeness, accuracy, consistency, traceability, and whether a particular expectation is mandatory or optional.

1.4 Align Project with Organizational Objectives

Organizational strategic priorities vary over time; successful projects are usually those that are aligned in a timely fashion with prioritized organizational objectives. Organizational generic strategic objectives include:

- Broad target cost leadership
- Broad target differentiation
- Narrow target focused low cost

| FIGURE 2-3 | Project House of Quality |

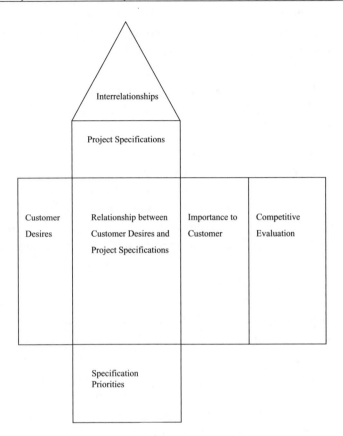

- Narrow target focused differentiation
- A combination of low cost and high quality.

"Broad" and "narrow" targets refer to the competitive scope of business strategies, while "focused" refers to the niche segments of quality features (differentiation) or low expenditure (cost leadership). Projects that are congruent with cost-cutting business priorities, quality-differentiation business priorities, or some combination of cost/quality tradeoffs are aligned with system priorities and avoid wasteful suboptimization. In effect, while a large number of creative projects may be feasible, only those projects that are aligned with organization system objectives are worth initiating.

Project managers, therefore, must be adept at identifying and defining high-level business-related priorities and system criteria for success. They must be able to clearly communicate the business justification for the project, articulate stakeholder expectations, and build consensus around the scope and value of the project. This project alignment process is the precursor to selecting core project team members.

1.5 Select Core Project Team Members

A high-level understanding of external customer expectations and alignment with organizational objectives are the requirements that must be achieved with project resources. It is now time to select the core team members using the PERA feedback tool.

A *project team* is a small number of people with complementary skills who are equally committed to a common purpose, goals, and working approach for which they hold themselves mutually accountable. Furthermore, a *core project team* differs from an ad hoc team in that the former ensures continuity of membership, preserving intact the resources of team talent from the beginning to the end of the project.

Ideally, these members will include a representative from most of the major disciplines that will be needed on the project; these core team members should be assigned for the entire project for the sake of continuity. When project team members are selected, consideration should be given to their individual personalities and the job responsibilities they will have, as well as to the interaction among various team members.

One common denominator on most projects is that time is at a premium. Therefore, it makes sense to ensure that project team members learn various time and stress management skills for coping with multiple priorities and deadlines. The project manager needs to ensure that team members have had similar project experience or be prepared to train them.

1.6 Determine Team Operating Principles

To maximize project performance, minimize conflicts, and generally make project work more enjoyable, the core team members will often determine team operating principles. *Operating principles* are guidelines and may be considered a charter for team interaction. A *team charter* is a document issued by the team outlining the conditions under which it is organized and defining its operational rights and privileges. It normally consists of a team values statement, a team mission statement, short- and long-term goals for team members, and a team operating agreement. Operating principles include how the team members will respect quality processes, how they will treat each other, team meeting planning and discipline, completion of work assignments, decision-making, and conflict resolution.

Some project teams that have worked together previously or that work for organizations with well-developed team methods may need only a few minutes to reaffirm existing team-operating principles. Others may need considerably more time to develop these guidelines. In any event, a team that functions well together is essential for achieving quality on any project. An explicitly endorsed team charter and explicit operating principles can be powerful guidelines and standards for quality team productivity.

Now that the needs of the customers are aligned with organizational priorities, key participants are selected, and operating principles are established, we turn our attention to the work process.

SECOND PROJECT QUALITY PILLAR: PROCESS IMPROVEMENT

Several process improvement activities should be performed during project quality initiation, as shown in Table 2-1. The project manager, sponsor, and core team need to adopt or develop a quality policy to ensure internal adherence to quality process improvement and external alignment with quality system standards. The main process improvement tasks in the project quality initiation stage are to flowchart the entire project process at a high level and to identify assumptions and risks. These tasks may be considered due diligence. Failure to perform either at this point is negligent and can dramatically increase the probability of undertaking a poor project or using a poor approach, both of which are quality problems. Next, the project manager, sponsor, and/or core team must establish a knowledge management process so that they can integrate past process lessons and consistently direct current knowledge acquisition processes.

2.1 Adopt or Develop Quality Policy

The core team needs to either adopt the quality policy of its parent organization (if it fits) or develop a project-specific quality policy if necessary. Typically, the *quality policy* identifies key objectives of products and services such as fitness for use, performance, safety, and dependability. Project managers, then, have the responsibility for defining, documenting, supporting, and communicating the quality policy of the organization and the project. Generally, part of the quality policy involves an internal and/or external audit program to determine if the activities and results of the quality system and the quality project are aligned.

2.2 Flowchart the Overall Project

Flowcharts are visual representations of how a process operates. At a minimum, flowcharts depict the starting and stopping points in a process, the activities performed, the decisions made, and the direction that materials, information, and people flow through the process.

A flowchart depicting the quality initiation stage is shown in Figure 2-2; flowcharts depicting each of the other quality stages will be shown in the next four chapters. The flowcharts developed during this quality initiation stage should be high level. The purpose is to show only enough detail so that the project team can determine the main approach its members will use to perform the project work, describe the work scope at a high level, and identify major project deliverables. This enables both the core team and the sponsor or client to sign a firm commitment so that both parties understand what will be accomplished. More detailed flowcharts are typically constructed during the planning stage.

2.3 Identify Assumptions and Risks

Quality on a project can suffer because either assumptions prove to be incorrect or known risk events happen. Identifying these potential problems at the outset can mitigate them. The sponsor and the core team should list the key assumptions they are making to ensure that both parties agree and to decrease the chance that faulty assumptions will lead to future quality problems. Both parties should then identify the major areas of the project in which they believe risk events are likely to occur.

Independently, the sponsor should define the level of risk he or she is willing to tolerate for each major area of the project and the core team members should estimate how much risk they believe exists in each of those areas. The goal of this simple analysis is to identify areas in which the core team believes the risk is higher than the sponsor is willing to tolerate. Identification of those gaps should lead to a different project approach, a better understanding of the chosen approach, or a higher risk tolerance on the sponsor's part. The documentation of assumptions and risks should be included in the project charter.

2.4 Establish Knowledge Management Processes

Once the quality policy exists, the overall project is flowcharted, and pertinent assumptions and risks are identified, the project manager and core team need a model to integrate the past lessons and to direct current knowledge acquisition activities. The model of a learning organization that manages its knowledge-based assets through structured project processes to achieve sustainable global competitive advantage is the motivational bedrock of project initiation.[3] Knowledge management tools that facilitate relevant organizational learning, as well as the production and sharing of knowledge, leverage the intellectual capital of organizations and accelerate the pace of directed innovation.[4]

The *plan-do-check-act (PDCA) model* depicted in Figure 2-4 is a knowledge management tool that provides direction. It starts with the "plan" step, during which a person will use his or her knowledge of a work process that needs to be improved, gather data about the current results of the process, select a potential improvement, and develop a plan to test the potential improvement. The "do" step occurs when the plan is piloted on a small scale and data are gathered to see if improvement occurs. The "check" step occurs when the process results from before and after the planned change are compared to determine the extent of improvement (if any). Finally, the "act" step can be to institutionalize the improvement if the results are good enough, to retest the improvement if the results are not yet good enough, and to return to the old process or a different test if the results are poor.

The PDCA can be used at several levels within a project. First, it can be used to incorporate improvements developed on previous projects into the new project right from the start. Second, it can be used to test new ideas

FIGURE 2-4   Plan-Do-Check-Act (PDCA) Model

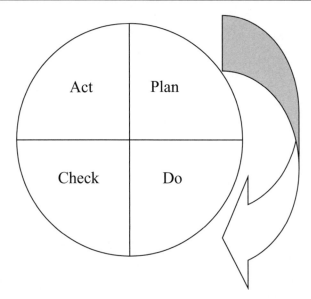

at one project stage and incorporate them into a future stage. Finally, any project participant, on any process, at any point, can use a PDCA to attempt improvement.

During this first stage of quality initiation, a project manager should plan the audits that will be used to help improve the project process. Also during this stage, the knowledge management goals should be stated. These should start with incorporating lessons from previous projects into the early plan for the current project.

Effective process improvement requires the third project quality pillar: fact-based management.

THIRD PROJECT QUALITY PILLAR: FACT-BASED MANAGEMENT

Fact-based management during quality initiation involves several activities, as shown in Table 2-1. Data will need to be collected and analyzed, producing accurate information or facts to be used for decision-making during each project stage. The first step is to reaffirm the norm of fact-based decisions and the use of quality tools to ensure reliable acquisition and use of data. The primary data to be used during the initiation stage consist of lessons learned both from previous projects and the quality initiation stage of the current project.

3.1 Agree to Make Fact-based Decisions

An important decision-making value is the individual and collective commitment to make *fact-based decisions* regarding project quality. To make fact-based decisions, the prevailing project norm means that the organization and the team must have the resources (including time) to determine what is factually true and the processes to ensure that members give voice to the truth.

To appreciate the importance of fact-based decisions with regard to project quality, it is useful to consider the alternatives of deciding without facts. Since the discovery process in uncovering facts is often prolonged, requiring mastery of many quality tools, many project managers and teams are tempted to make fiat-based rather that fact-based decisions. Irresponsible project managers and project teams may be tempted to make key decisions about project quality on the bases of subjective whims, unfounded intuitive hunches, or uncritical groupthink dynamics. Fact-based decisions, however, provide a self-correcting, truthful foundation for project quality decisions and provide an objective basis for project learning in the future.

3.2 Identify Lessons Learned from the Past

While individuals and teams can "reinvent the wheel" with each new project, it is preferable for an organization to store, retrieve, and transfer project lessons learned from its past. The lessons can be categorized into tacit and explicit knowledge. Tacit lessons are subjective and practical while explicit lessons are objective and theoretical. Sometimes project managers with seniority in an organization and professional project experience can orally transfer their knowledge to newer project managers. Often, however, with rapid economic and organizational changes, the relevant lessons learned from the past may require other *modes of project knowledge conversion*, as indicated in Figure 2-5.

Socialization is an informal process of sharing tacit experience. For example, when project team members apprentice to a project manager, they learn through observation, imitation, and practice since language may not be a sufficient vehicle for transmission. Externalization is a formal process of articulating tacit knowledge into explicit concepts. In spoken and written words, tacit knowledge may take the form of metaphors, concepts, or equations in project management manuals or in handbooks for specific tasks or industries. Internalization is the absorption of explicit knowledge into tacit knowledge though oral transmission of project lessons, systems document processes, or simulations. Combination is the process of systemizing explicit concepts into new explicit knowledge by analyzing, categorizing, and reconfiguring information (e.g., university project management education

FIGURE 2-5   Modes of Project Knowledge Conversion

	To	
	Tacit Knowledge	Explicit Knowledge
From — Tacit Knowledge	Socialization	Externalization
Explicit Knowledge	Internalization	Combination

using databases and computer networks to supplement the lectures of professors). Each of these four modes of knowledge conversion has its place. A wise project manager will attempt to use all four to accelerate project learning.

Two traditions for identifying lessons learned are the professional (or content "what" experts) and the organizational (or process "how" experts). While there are tensions between the two, when both cooperate to reinforce each other, new project managers can rely on prior professional lessons about what should be done and prior organizational lessons about how it should be done to be operationally successful in this organization.

3.3 Collect and Share Project Quality Initiation Lessons Learned

One of the deliverables that is expected at the end of each project stage is a set of lessons learned from that stage. These lessons learned should be used both to improve future stages of this same project and to be part of the end-of-project documentation that will improve future projects. In addition, through sharing, the project participants contribute to sustaining a learning organization.

There are many ways to collect lessons learned. One simple method is called the *plus-delta*, an example of which is shown in Figure 2-6. This can be easily facilitated and can be used on a portion of a project as simple as a meeting or as complex as the entire project. The facilitator draws a large "T" on a flip chart with a plus sign (representing positive things) over the left crossbar and a large triangle (representing things to change) over the right crossbar. Then project participants state what they thought was positive and should be repeated during the future of this project or on future projects as well as negative things they feel should be changed in the future. The

FIGURE 2-6   Plus Delta Project Evaluation

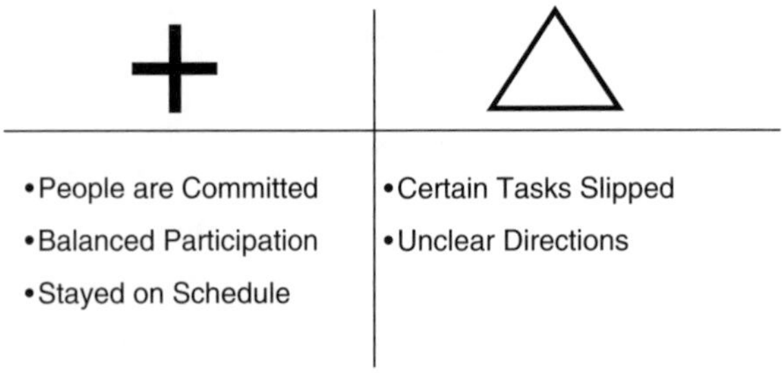

- People are Committed
- Balanced Participation
- Stayed on Schedule

- Certain Tasks Slipped
- Unclear Directions

facilitator writes these on the flip chart. The wise leader will attempt to find obvious ways to use these ideas so participants will feel that their ideas are important. This motivates the participants and provides a natural transition into our last quality pillar: empowered performance.

FOURTH PROJECT QUALITY PILLAR: EMPOWERED PERFORMANCE

Project quality initiation ultimately requires empowered, committed, and principled performance from every project participant to persevere over time rather than prematurely abandon projects. To determine the likely sustainability of commitment in a particular organization, it is advisable to assess its ethical work culture values. Organizations in which fear and distrust prevail will eventually disempower project managers and undermine the best quality projects. Unless the integrity capacity of the organization can be developed, the next two steps of this phase—project selection and commitment—may not mean very much. If, however, the work culture is sufficiently morally developed to respect principled project selection choices and will honor formal commitments to projects, it is worthwhile to proceed.

4.1 Develop Ethical Work Culture Values

Developing an *ethical work culture* that values responsible project initiation ensures support for project quality. Individuals, teams, and organizations that value different levels of moral development enhance or inhibit successful project quality. Those that morally prize the direct and/or indirect use of force as the determinant of workplace norms value a context of manipulation. This allows fear to determine project quality levels. Those that morally prize conformity to internal operating procedures and external legal/regulatory authority value a context of compliance. This implicitly endorses management by conventional authorities rather than management by fact. Finally, those that morally prize democratic participation and universal principles value a context of committed integrity capacity.[5] This allows reasonable, fact-based evidence rather than power or conventional authority to determine project quality levels.

Many unsound projects are initiated because their work culture contexts are so politicized that reasonable project quality standards have little or no chance of being maintained if they threaten powerful interests inside or outside the organization. Successful project quality is likely to be sustained only if the work context is above the compliance level of moral development. This occurs because if there is no internalized commitment to project quality, when external champions depart or project quality enforcement pressures wane, standards will be rapidly abandoned.

A tool for determining the level of individual and collective moral development is the *Ethical Work Culture Assessment* (EWCA) presented in Appendix B. Using this tool will provide the project sponsor, the project manger, and the project team a measure of the level of moral development and work culture support for sustaining sound projects aligned with organizational objectives. Either it will indicate a high level of moral development or it will identify areas that need to be improved to move toward the desired level.

4.2 Select Project

Once the project sponsor has determined that the project is a good fit for the organization, it is time to formally select the project. The selection means that the organization will officially support the project. While this is obviously important, it is still a precursor to the charter signing by the project manager, sponsor, and core team members.

The sources of project identification are diverse, ranging from personal creativity to impersonal system-generated tasks that require attention and resources. Project quality will best be achieved, however, by selecting projects that are aligned with the strategic priorities of the organization, are statistically warranted, and are likely to secure the needed commitment of capable participants to bring projects to successful closure. "Pet" projects of powerful sponsors that are not strategically aligned, not statistically warranted, and/or lack the requisite critical mass of committed support from capable participants should be screened out of consideration.

Many methods are used to select projects. Some are much more involved than others. Our advice is to choose a method that is sufficient to include factors that are important to your organization but that is no more complicated than necessary.

4.3 Formally Commit to Project

The final element of project initiation is the personal public commitment to the project. The quality initiation stage-ending document is either a contract or a letter of intent for an external project. The equivalent of a contract for an internal project is a signed agreement between a project sponsor and a project core team that is called a charter. Individual core team members will sign the charter to signify their individual commitment to the project. An example of a *project charter* is shown in Figure 2-7.

| FIGURE 2-7 | Project Charter |

Project Name: _____ Date: _____

	Name	Signature	Responsibilities
Project Sponsor:	_____	_____	_____
Project Manager:	_____	_____	_____
Core Team Members:	_____	_____	_____
	_____	_____	_____
	_____	_____	_____

Business Need: _____

Project Purpose: _____

Scope Overview: _____

Project Deliverables: _____

Customer Acceptance Criteria: _____

Team Operating Principles: _____

Lessons Learned to be Used: _____

Assumptions & Risks: _____

Charters are very powerful. Project charters are used to:
- Clarify the project purpose
- Set clear project goals
- Develop teamwork
- Develop common understanding, trust, communication, and commitment between the sponsor and the core team
- Avoid situations in which the core team is unsure if management will accept an action or decision
- Avoid situations in which the sponsor unilaterally changes the original agreement.

The process of developing and ratifying a project charter starts with one party (the sponsor or core team) writing a rough draft. A short draft encourages all involved parties to read, understand, discuss, and negotiate. The other party questions everything for both understanding and agreement. Eventually, both the sponsor and core team sign the project charter.

The project charter should then have the force of a contract. That means that both the sponsor and the core team feel bound by it and will try their level best to live up to the terms of the charter. Like a contract, the charter can be modified only if both parties agree.

Once the key participants have publicly and personally committed to the project, its chances for quality problems have certainly decreased. This commitment is the stage-ending deliverable that transitions the project from project quality initiation into project quality planning.

Up to this point, the project work activities have consisted of creating high-level understanding—just detailed enough for all parties to reach commitment. Planning takes time, and therefore costs money. As such, we only want to sink time and money into projects to which all key stakeholders are committed. During the project quality initiation stage, a few potential projects that cannot obtain commitment from all key stakeholders will be abandoned. Now that we are ready to proceed into project quality planning, with all the time and cost it entails, we are assured that the project we are planning has solid prospects.

NOTES

1. Project Management Institute Standards Committee, *A Guide to the Project Management Body of Knowledge* (*PMBOK® Guide*) (Upper Darby, PA: Project Management Institute, 2000), p. 5.
2. James R. Evans and William M. Lindsay, *The Management and Control of Quality*, 5th edition (Cincinnati, OH: South-Western Publishing, 2002).
3. Karl E. Sveiby, *The New Organizational Wealth: Managing and Measuring Knowledge-based Assets* (San Francisco, CA: Berrett-Koehler, 1997).
4. Thomas A. Stewart, *Intellectual Capital: The New Wealth of Organizations* (New York: Currency, 1997).
5. Joseph A. Petrick and John F. Quinn, "The Integrity Capacity Construct and Moral Progress in Business," *Journal of Business Ethics* 23 (2000), 3-18.

CHAPTER 3

Project Quality Planning

Planning is defined in the *PMBOK® Guide* as "the process in which defining and refining objectives and selecting the best of the alternative courses of action to attain the objectives that the project was undertaken to address are performed."[1] The quality planning stage begins with a commitment and authorization to proceed on a project, and ends with the kick-off meeting of project participants that signals the start of project execution. Quality planning follows quality initiation as the second stage in the five-stage project quality process model shown in Figure 3-1.

As is true of all five stages, the management activities will be much more involved on some projects than on others. Large, complex, unfamiliar projects will require more in-depth planning than smaller, simpler, more familiar projects. The typical quality planning activities required are depicted in the flowchart in Figure 3-2.

Project quality pillars, project activities, and project tools facilitate the movement from the signed authorization to proceed to the point at which all project stakeholders commit to the project plan. Table 3-1 categorizes the project quality pillars, activities, and tools for the quality planning stage into a project factors table.

This chapter is structured to follow the order of the project quality pillars and their sequenced activities shown in Table 3-1. The first number of the listed activities corresponds to the appropriate quality pillar, e.g., Activity 1.1 is associated with the first pillar and Activity 2.1 is associated with the second pillar. The second number refers to the typical approximate chronological sequence of its execution within the pillar's domain, although this sequential order may well vary with different projects, organizations, or industries. For example, 1.1 Determine Customer Satisfaction Standards normally comes before 1.3 Determine Levels of Decision-Making Authority.

FIRST PROJECT QUALITY PILLAR: CUSTOMER SATISFACTION

As is true in each project stage, several activities relating to customer satisfaction should be completed. Project quality planning begins with a determina-

FIGURE 3-1   Five-Stage Project Quality Process Model

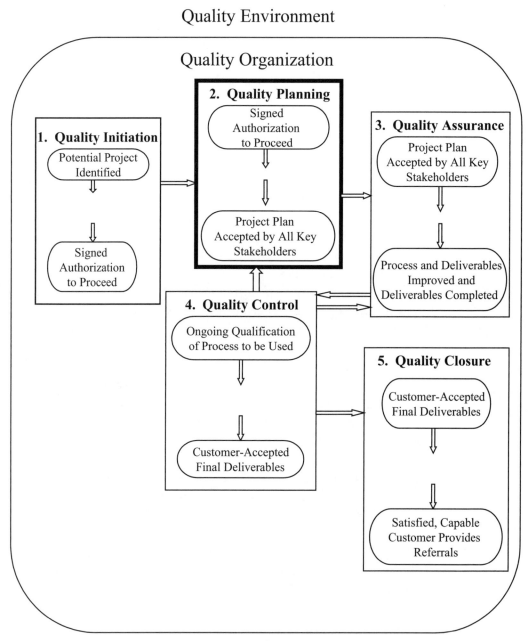

Quality Environment

Quality Organization

Quality Context = Quality Organization + Quality Environment

FIGURE 3-2 Project Quality Planning Flowchart

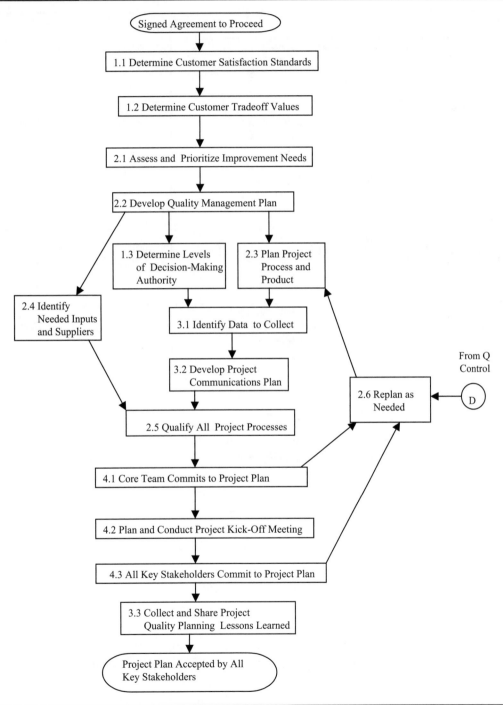

TABLE 3-1 Project Quality Planning Factors Table

Pillars	Activities	Tools
1. Customer Satisfaction	1.1 Determine Customer Satisfaction Standards 1.2 Determine Customer Tradeoff Values 1.3 Determine Levels of Decision-Making Authority	Customer Standards Matrix Customer Tradeoff Values Matrix Project Decision Responsibility Matrix
2. Process Improvement	2.1 Assess and Prioritize Improvement Needs 2.2 Develop Project Quality Management Plan 2.3 Plan Project Process and Product 2.4 Identify Needed Inputs and Suppliers 2.5 Qualify All Project Processes 2.6 Replan As Needed	Cause and Efrfect Diagram Benchmarking, Cost/Benefit Analysis JAD Sessions, Concurrent Engineering SIPOC Model Process Quantification Levels
3. Fact-Based Management	3.1 Identify Data to Collect 3.2 Develop Project Communications Plan 3.3 Collect and Share Project Quality Planning Lessons Learned	Data and Measurement Matrix PDCA Plus Delta Model
4. Empowered Performance	4.1 Core Team Commits to Project Plan 4.2 Plan and Conduct Project Kick-Off Meeting 4.3 All Key Stakeholders Commit to Project Plan	Kick-Off Meeting Agenda and Minutes

tion of customer satisfaction standards and tradeoff values. If the project sponsor, manager, and core team fully understand these two customer desires, the chances of completing a high-quality project are much greater. The project sponsor, project manager, and core team then need to determine the levels of decision-making authority to avoid strategic confusion and operational conflict.

1.1 Determine Customer Satisfaction Standards

Since the customer ultimately judges the quality of the project output, the project team needs to understand the customer's standards of satisfaction. The best way to gain this understanding is to ask external and internal customers directly.

The team then must use this knowledge to develop the specifications for the project output as well as the process steps that will be used to meet the desired standard. The *customer standards matrix* depicted in Figure 3-3 facilitates this process by providing information in customer criteria for

FIGURE 3-3   Customer Standards Matrix

Criteria	Measure	Standard

quality project acceptance, methods for measuring criteria compliance, and detailed information on reasonable targets for project quality attainment.

The core team can ask the customer(s) to specify the important criteria upon which they will judge the quality of the project. For each criterion, the customers should then describe how they will measure that criterion and what the standard for satisfaction will be. Quite frequently the customers will not know these standards in advance, so a facilitated discussion with the core team may be necessary. The dimensions of quality in manufacturing and service listed in Chapter 1 may be helpful in getting the customer to determine what criteria are important. This essential step of identifying customer satisfaction standards is frequently left out, and the result is that the project team is left guessing how the customer will judge the quality of the project output. How can someone be confident of producing high-quality output if he or she does not know what the customer wants?

1.2 Determine Customer Tradeoff Values

Customers will have priorities among the various project objectives. The project manager should ask each external and internal customer to prioritize among the project objectives of cost, schedule, quality, scope, contribution to the organization, and contribution to society. Using the *project customer tradeoff matrix* shown in Figure 3-4, the manager should ask which objectives

FIGURE 3-4 Project Customer Tradeoff Matrix

	Enhance	Maintain	Sacrifice
Cost			
Schedule			
Quality			
Scope			
Contribution to Organization			
Contribution to Society			

should be enhanced if possible, which objective(s) must be maintained, and which objectives can be sacrificed if needed.

Many customers will not initially be willing to admit that they would consider sacrificing any objective. Therefore, the project manager needs to have a frank discussion to impress upon each customer group that the project team will do its best to achieve all six objectives. However, during the course of the project, decisions will invariably need to be made under unanticipated contingencies, and understanding which objectives are relatively more important will help the project manager make the kind of decisions that the majority of customers will likely accept.

The project manager and sponsor must also decide which customer groups are relatively more important if priorities conflict. Typically, external paying customers and top management are considered quite important. If significant differences in priorities surface among various customer groups, more consideration may be necessary at this point.

Additionally, the project team should consider possible preplanned product improvements—particularly if the sponsor or a majority of customers choose to enhance performance. In any event, if the project team starts thinking about product and process improvements early, both current and future projects are likely to benefit. The objectives selected for enhancement by the customers (cost, schedule, quality, scope, contribution to the organization, and/or contribution to society) should direct the team's thinking as the members continually strive for improvement.

1.3 Determine Levels of Decision-making Authority

One frequent cause of quality problems is that project participants do not know who is allowed to make certain decisions. This problem can be minimized if the proper decision-makers have the time, information, and skill to make decisions and understand their respective roles. The *project decision responsibility matrix* in Figure 3-5 is a tool for clarifying three decision-making factors relating to specific issues: (1) who must be informed, (2) who is authorized to make recommendations, and (3) who is authorized to decide.

For each issue that must be decided, responsibility for making the recommendations and being informed should be noted. A recommended approach is to have one primary decision-maker per issue (others may recommend), with the project manager at least informed about virtually every issue. While all project participants have roles, the project manager is ultimately responsible for quality and must know what is happening.

FIGURE 3-5   Project Decision Responsibility Matrix

Issues to be Decided	Project Participant's Role							
	Customer	Sponsor	Project Manager	Functional Manager (specify)	Technical Lead (specify)	Core Team	Individual Team Member	Other Stakeholder (specify)
A	D	I	R					
B			I			D		
C			D					
D			I	D				
etc.								

R = Recommend
D = Decide
I = Must Inform

SECOND PROJECT QUALITY PILLAR: PROCESS IMPROVEMENT

Carpenters are told to measure twice and cut once to avoid making mistakes. Likewise, project managers are encouraged to put extra effort into planning their work processes to avoid quality problems later. Many process-related issues should be settled during the project quality planning stage.

An initial assessment and prioritization of process improvement needs based on root cause analysis are necessary to determine whether and to what extent incremental improvement, competitive parity, or breakthrough dominance are warranted with different processes. Then, a quality management plan needs to be developed to comprehensively address the diagnosed problems and opportunities. Next, both the process and the product of the project should be planned simultaneously. The customer value supply chain needs to be identified, along with the inputs each will provide. All processes need to be qualified. Finally, planning is iterative and a great deal of replanning typically is needed. Good process planning goes a long way toward good project quality.

2.1 Assess and Prioritize Process Improvement Needs

When assessing a process, one of the most important things to understand is the process variation. Variation is often the reason for poor quality. The project manager needs to understand not only what kind of variation exists, but also what the causes are of that variation. Armed with this knowledge, a smart manager can reduce or eliminate the problem variation.

Variation in project process output and other problems can occur for many reasons, including materials, machines, methods, people, measurement, and environment. The first goal of project process improvement is a correct diagnosis of the root causes of problems and a relative ranking of their severity or urgency for prioritized attention. Otherwise, project managers can end up treating symptoms rather than causes and attacking marginal rather than critical causes.

The *cause-and-effect diagram* (or *fishbone diagram*) in Figure 3-6 provides one way to identify process problems, their root causes, and their contributory causes. The fishbone diagram is constructed so that the fish head shows the problem, each major branch (machines) pointing into the main stem represents a possible cause, and minor branches (defective parts) pointing into the major branches are contributors to the cause of the process problem. The diagram, therefore, identifies the most likely causes of process problems

FIGURE 3-6   Cause and Effect Diagram

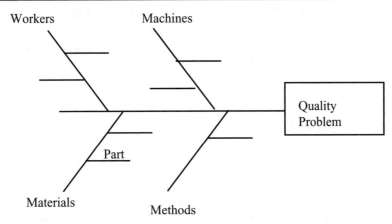

so that project participants are properly focused to collect and analyze relevant data to prove which of the possible causes are actual causes.

In addition, the involvement of the project team and other customers in the brainstorming input into this causal analysis usually heightens their awareness of problem causes and their sense of commitment to resolutions. Other tools such as regression analysis and statistical correlation analysis will identify positive, negative, or neutral correlations between and among the variables.

Among the key factors in this step is determining what type of improvement is sought for each project work process. Given different issues with different priorities, the project team must decide whether proposed resolutions will aim at incremental improvement, competitive parity, or breakthrough dominance. The team also needs to determine which process improvements should come first.

2.2 Develop Project Quality Management Plan

After assessing and prioritizing the process improvement needs, the *quality management plan* is developed. This master plan links all prioritized process needs and resource requirements to strategic priorities and ensures that all project participants have clear goals and delineated responsibilities. All process and product standards are framed in concrete, measurable terms through operational definitions such as time deadlines, budgetary cost limits, and product tolerance specifications. Tentative fund allocation decisions are displayed and procedures for plan review and alteration are specified. The project quality management plan answers the questions:

- **What** projects are being funded?
- **Where** and **why** are they strategically prioritized?
- **When** are they scheduled to begin and end?
- **How** (in terms of timelined process steps and detailed procedures to meet standards) will they be tracked and executed?
- **Who** is responsible for performing work during each phase of each project and for sharing information?
- **How many** resources (financial and non-financial) are required?

In addition, the plan requires emergent feedback loops so that operations in the field can provide corrective information while the plan is being deployed and reformulated.

In developing this master plan, the project management team relies on internal and external benchmarking of products/services and processes. Longitudinal data and trends of prior internal organizational performance standards (along with horizontal comparative data and trends from external sources) ensure that the master plan has the necessary benchmark information to enact competitive parity or breakthrough dominance improvements based on factual input.

Process benchmarking, for example, starts with identifying specific processes that the team wants to improve. Next, the project team identifies an organization that is particularly strong in that process. Then, the project manager contacts a manager at the organization against which he or she wants to benchmark. The project team prepares focused process questions in advance and someone from the project team makes a site visit to observe the process in action and interview key employees. The project team then analyzes the data by identifying gaps between what they are presently capable of doing and what the benchmark organization is doing. Finally, the team decides what aspects of the benchmarked process they can use as is in their project, what must be adapted for the project, and what is not relevant or achievable at this time.

Once the project team has finished benchmarking, they can also use cost/benefit analysis to determine if a proposed approach makes economic sense. The cost/benefit analysis should consider the operating conditions and other environmental factors that the user of the project deliverables will experience.

2.3 Plan Project Process and Product

Now that process improvement needs are prioritized and a quality management plan is developed, it is time to jointly plan project products and the work processes that are needed to create them. For planning project processes and products effectively in a quality manner, two tools are useful: joint application design sessions and concurrent engineering.

A *joint application design* (JAD) session is used to ensure that all customer desires are identified and prioritized, as well as to develop the technical approach, estimate the time required to develop the technical approach, and estimate the time required to develop each prioritized project feature. Prototypes are sometimes used in JAD sessions with the users to refine the requirements and to identify missing functionality. This step is used when customers lack experience and need concrete examples to extract business requirements.

A JAD session consists of two parts. During the first part of the JAD session, the customers identify possible enhancements and prioritize each while developers achieve enough understanding of the functional requirements that they can (during the second part of the JAD session) estimate the time required. The first part of the JAD session starts with the facilitator reading each potential enhancement aloud. For the enhancements that have long descriptions, the facilitator encourages a knowledgeable participant to paraphrase the description. Once the description is read or paraphrased, five questions are answered for each feature and eventually prioritized:

1. What do we not understand about this request?
2. What is the business reason for this request?
3. What is the impact of not doing this enhancement?
4. What action items need to be accomplished?
5. What impact does this have on other parts of the project?

During the second portion of the JAD session, only the developers go through the proposal features. For each feature, the project manager identifies who must estimate the time required. If individuals need additional information, probable estimates are presented until more information is obtained. People must have enough time to envision how much work is involved without getting trapped into detailed development discussions.

The other useful tool in project product and process improvement is concurrent engineering. *Concurrent engineering* is a process in which all key project participants involved in bringing a product/service to market are continuously involved with that product/service development from conception through sales. This simultaneous rather than sequential process shortens product development cycles, lowers costs, reduces rework, and generally addresses quality issues at an early stage.

Project teams that take the time to simultaneously plan both the project outputs and the processes to create them greatly lessen the chances of unpleasant surprises (quality problems) later.

2.4 Identify Needed Inputs and Suppliers

Now that the processes needed to produce the project output are understood, it is time to identify the inputs that are needed. In addition to the house of quality, the *supplier-input-process-output-customer (SIPOC) model* is a tool that can be used to improve the project process by clearly identifying relationships among suppliers, inputs, processes, outputs, and customers. An example is shown in Figure 3-7.

The SIPOC is a visual guide to help a project team work backwards from customers to identify all the project customers (C), including unintended stakeholders who are impacted by the project. The SIPOC next guides the team in identifying what product, service, and information outputs (O) each customer wants to receive (or receives inadvertently) and the satisfaction standards that customers demand from each output of the project. The third item the team uses the SIPOC to identify is the set of process (P) actions the project team needs to take and the standards that must be set in order to create those identified outputs. Flowcharts, introduced in Chapter 2, are often used to illustrate this process portion of a SIPOC.

The fourth item that teams use the SIPOC to identify is the set of information, workers, material, or other inputs (I) needed to meet the process standards. Finally, the SIPOC guides the team in identifying the suppliers (S) of the desired inputs. A list of quality suppliers can then be generated to sustain long-term quality improvement partnerships with solid domestic and global suppliers.

2.5 Qualify All Project Processes

Organizations that produce excellent quality outputs insist on using excellent processes to produce their outputs. One method of ensuring that

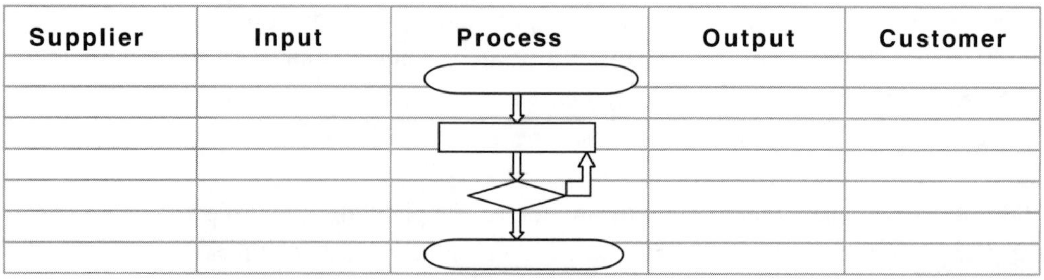

FIGURE 3-7   Supplier-Input-Process-Output-Customer (SIPOC) Model

only excellent processes are used is to qualify each process. Process qualification levels from spontaneous to optimized status have already been addressed in Chapter 1. However, once strategic alignment and process improvement priorities have been decided, the ongoing qualification of all project processes will determine the rate of efficiency and effectiveness improvement over the course of the project.

2.6 Replan As Needed

Since the master plan requires ongoing feedback from project implementers, the likelihood for replanning is high. The master plan provides for three feedback channels:

1. Outside to inside
2. Inside from top to bottom
3. Inside from bottom to top.

The openness to feedback during project formulation and implementation from the outside (customers and others) allows external benchmarking data and external stakeholder voices to have an impact on project replanning. The top to bottom feedback is customary in hierarchical organizations. Feedback from bottom to top allows all participants to have a voice often and uses information technology. An example is project operators using laptops to e-mail replanning suggestions to the project manager, sponsor, or even CEO. The more sources of feedback that are considered, generally the better the replanning effort becomes. Prudent project managers and teams learn how to sort though vast quantities of data to quickly find the most useful information for their replanning.

THIRD PROJECT QUALITY PILLAR: FACT-BASED MANAGEMENT

While all the project planning and replanning are occurring, several issues concerning data must be resolved. The team needs to identify the data that must be collected, develop a project communications plan, and capture lessons learned for project participants.

3.1 Identify Data to Collect

A fundamental part of making fact-based decisions is gathering data that can be compiled and interpreted into the facts needed to make sound project decisions. The science of determining what data to collect, how to define the data, how to collect the data, how to analyze the data, and how to use the data in decision-making is called *metrics*. In an effective metrics system:

- Variables are operationally defined
- Normal variation is distinguished from abnormal variation
- All project stakeholders have a common understanding of project status
- Metrics are practical and easy to obtain
- Metrics are collected at regular intervals
- Management accepts metrics.

The SIPOC model is a useful starting place to determine some of the needed metrics, such as customer satisfaction standards (see Figure 3-7). This can be used both for identifying measures to collect and for setting goals. Projects are conducted either within one company or among multiple companies. In either event, the project metrics need to align with those of the parent organization and any other organizations involved.

Another tool that can be used to determine what data needs to be collected is the *data and measurement matrix*, shown in Figure 3-8. While this simple tool can be used to help determine what needs to be collected with regard to cost, schedule, scope, and quality, the emphasis here is on quality. The project team (often in conjunction with the customer) determines the various quality factors (such as errors in a software project) that need to be monitored. Once the quality factors have been identified, the team needs to state clearly what data are needed to indicate the level of expected quality (such as how many errors one could expect at each project checkpoint if the project is proceeding according to plan). Then the team needs to determine what data will be collected to determine the actual status of each quality factor. Finally, the team should determine the monitoring activities—who will collect the data and how.

3.2 Develop Project Communications Plan

A major source of quality problems on projects is faulty communications. Keeping all project participants accurately informed leverages the firm's resources and sustains momentum. To decrease the likelihood of making mistakes, it is imperative for a project core team to develop a comprehensive *project communications plan*—and to use it. Most people receive far more communications than they need, so the answer is not more information, but more useful, specific information.

Project communications is an excellent opportunity to use the PDCA model (see Figure 2-4). The project core team first plans **(Plan)** who needs to know what information, how often they need it, and their preferred

FIGURE 3-8   Data and Measurement Matrix

	EXPECTED DATA NEEDED	ACTUAL DATA TO BE COLLECTED	MONITORING ACTIVITIES
Cost			
Schedule			
Scope			
Quality Factor 1			
Factor 2			
...			
Factor N			

information format. Next, the team uses **(Do)** the communications plan. Very quickly and repeatedly, the team should seek feedback **(Check)** on the quality and completeness of the information being transmitted through the communications plan, using information technology wherever feasible. Finally, the team should **Act** upon the feedback by improving the communications plan.

3.3 Collect and Share Project Quality Planning Stage Lessons Learned

Periodically throughout the project—at least at the end of each stage—lessons learned should be captured to help conduct future stages better. The plus delta tool can be used for collecting the lessons learned at the end of the planning stage. Lessons learned should then be used to improve future stages of the current project and other projects, and through sharing, contribute to the organization's learning capacity.

FOURTH PROJECT QUALITY PILLAR: EMPOWERED PERFORMANCE

All four project quality pillars are important. The first three (customer satisfaction, process improvement, and fact-based management) reinforce and are supported by the fourth (empowered performance). In other words, doing a good job on the first three helps empower individual performance, and outstanding individual performance through empowerment really drives successful accomplishment of the other three pillars.

The determinants of empowered performance that must be achieved at the end of the project quality planning stage are the commitments of the core team and all project stakeholders to accept the detailed project plan. Once the core team members review the entire plan and determine that they want to commit to it, they will informally sell the project plan to the diverse project stakeholders. Nevertheless, a formal project kick-off meeting is a useful public ritual to answer organizational concern and to solidify organizational support.

4.1 Core Team Commits to Project Plan

Empowered performance will not energize the project unless and until the core team commits to the project plan. If the team members believe that improvements are needed in the project plan, this is the time to make them. There will always be replanning, both to elaborate with more details and to respond to customer changes or other changing conditions. Nevertheless, the core team members and the project manager must each personally commit to the detailed project plan before they can convince the sponsor, customers, and other stakeholders to commit. Many wise sponsors have approved project plans that they felt were less than ideal because the project manager and the core team were so passionately committed. The sponsor's trust in such situations is usually rewarded because the team finds ways to overcome obstacles.

While we prefer excellence in everything, we would rather have a good plan and a passionate team than an excellent plan and a compliant team. One note of caution: Even the most committed team cannot usually overcome a seriously flawed plan. All participants need to use good judgment along with their enthusiasm.

4.2 Plan and Conduct the Project Kick-Off Meeting

The *kick-off meeting* serves to transition from planning the project to executing it. The core team that has performed much of the planning now shares with all the people who will perform the project work, including suppliers, ad-hoc workers, etc. The project work should be described in broad terms and then the participants should have the opportunity to ask as many questions as they like.

The PDCA model (shown previously in Figure 2-4) can be used to study and improve the kick-off meeting process just as it can be used to improve many other work processes.

The "Plan" step includes reviewing the project charter, considering carefully who needs to attend (maybe some participants only need to attend a portion of the meeting), and creating a detailed agenda for the kick-off meeting. (See Figure 3-9 for an example of a kick-off meeting agenda.)

The "Do" step is conducting the kick-off meeting. There are three types of goals for the kick-off meeting participants: building relationships, understanding tasks, and learning. The typical kick-off meeting has a structured order and starts with a quick review of the agenda (asking whether anything should be modified). Topics are covered one at a time. The core team member who was assigned the task of presenting a topic will do so, recommending an approach, and the core team members will answer questions until a collective decision is reached. Finally, as the meeting comes to a close, the project manger will summarize decisions made and assign action items for each participant.

The "Check" step incorporates the meeting evaluation stage. The last item on the agenda for meetings should be an evaluation. The plus delta model shown previously in Figure 2-6 is a useful technique for improving meetings of any kind.

The "Act" step is following up on all the decisions made in the kick-off meeting and striving to improve future performance on both current and future projects. To ensure that all agreed-upon tasks are completed, good meeting minutes should be written up and distributed promptly so participants can create their detailed project plans. (See Figure 3-10 for an example of a kick-off meeting minutes template.)

4.3 All Key Project Stakeholders Commit to Project Plan

To increase the likelihood of commitment by all key project stakeholders, it is helpful to anticipate and have detailed answers to the following questions:

- Why this project?
- Why now?
- Are financial resources adequate?
- Are human resources adequate?

FIGURE 3-9   Project Kick-Off Meeting Agenda

Project _____

Attendees _____ _____ _____ _____

_____ _____ _____ _____

Date _____ Time _____ Place _____

When	What	Who	Expected Outcome
8:00	Agenda Review	Project Manager (PM)	Understanding
8:10	Project Introduction	Sponsor, PM	Realize Importance
8:30	Participant Introduction	All	Meet Each Other
. . .	Work Expectations	PM, Core Team	Agreement
. . .	Project Goals	PM	Understanding
. . .	Customer Satisfaction Stds	Sponsor	Understanding
. . .	Project Plan and Status	PM	Understanding
. . .	Quality and Communications Plans	Core Team	Introduction
. . .	Questions and Answers	PM, Core Team	Understanding
. . .	Project Plan Revisited	PM	Commitment
. . .	Action Items	PM	Agreement
. . .	Meeting Evaluation	PM	Improvement

- How thoroughly do you understand the customers?
- How likely is it that customer requirements will change?
- How often and by how much will they change?
- Are appropriate data identified?
- Is the data gathering and analysis system adequate?
- Have customer rights been described?
- Are standards identified or developed by which the project will be judged?
- Are both deliverables and work processes to create them as simple as practical?
- How does the rest of the organization benefit from this project's success?
- How does society benefit from this project's success?

It is also helpful to share the plan (or portions of it) with many of the other (non-key) stakeholders who could potentially disrupt the project. The

---

FIGURE 3-10   Kick-Off Meeting Minutes Template

_____ Project Team   Date: Time:

Members present:

Information Shared

Decisions Made   (Alternatives Considered)

Issues to be Addressed Later

Action Item   Persons Responsible   Completion Date

Meeting Evaluation

project manager, sponsor, and core team should consider who might be an ally to the project if courted and who could become an enemy if not courted. Then they should develop and execute a strategy of trying to win over the various groups.

Often, a particular stakeholder is interested primarily in only one small aspect of the project. When that is the case, sharing why the project is important and showing a willingness to make adjustments (if practical) can help a minor stakeholder become less negative about the project. An ongoing dialogue may be necessary throughout the project with many diverse stakeholder groups. Nevertheless, project quality planning is complete when all key project stakeholders have agreed to the project plan.

NOTES

1. Project Management Institute Standards Committee, *A Guide to the Project Management Body of Knowledge* (*PMBOK® Guide*) (Upper Darby, PA: Project Management Institute, 2000), p. 30.

Project Quality Assurance

After project quality planning, both project quality assurance and project quality control begin. Chapter 4 covers project quality assurance and Chapter 5 covers project quality control.

Quality assurance can be defined as "all the planned and systematic activities implemented within the quality system to provide confidence that the project will satisfy the relevant quality standards."[1] For the sake of clarity, we define quality assurance activities to start when the key project stakeholders approve the project plan and the focus of activities shifts from strictly planning to mostly execution. Quality assurance activities continue until the final project deliverables are complete. Quality assurance follows quality planning as the third stage in the five-stage project quality process model and runs largely parallel with project quality control (see Figure 4-1).

The project quality assurance stage as the third stage and the project quality control stage as the fourth stage have unique and dynamic interactions centered around the process improvement and fact-based management tasks. As is true for all the process stages, the level of detail needed during the project quality assurance stage can vary significantly from one project to another. Typical project quality assurance activities are shown in Figure 4-2. The flow of information both from and to the fourth stage (the project quality control stage) is also depicted on this flowchart.

The pillars, activities, and tools that accomplish the tasks during this process stage are listed in Table 4-1.

This chapter is structured to follow the order of the project quality pillars and their sequenced activities depicted in Table 4-1. The first number of the listed activities corresponds to the appropriate quality pillar, e.g., Activity 1.1 is associated with the first pillar and Activity 2.1 is associated with the second pillar. The second number refers to the typical approximate chronological sequence of its execution within the pillar's domain, although this sequential order may well vary with different projects, organizations, or industries. For example, 2.1 Conduct Ongoing Review of Project Process Adequacy normally comes before 2.3 Improve Processes based on Data Analysis.

FIGURE 4-1 Project Quality Process Model

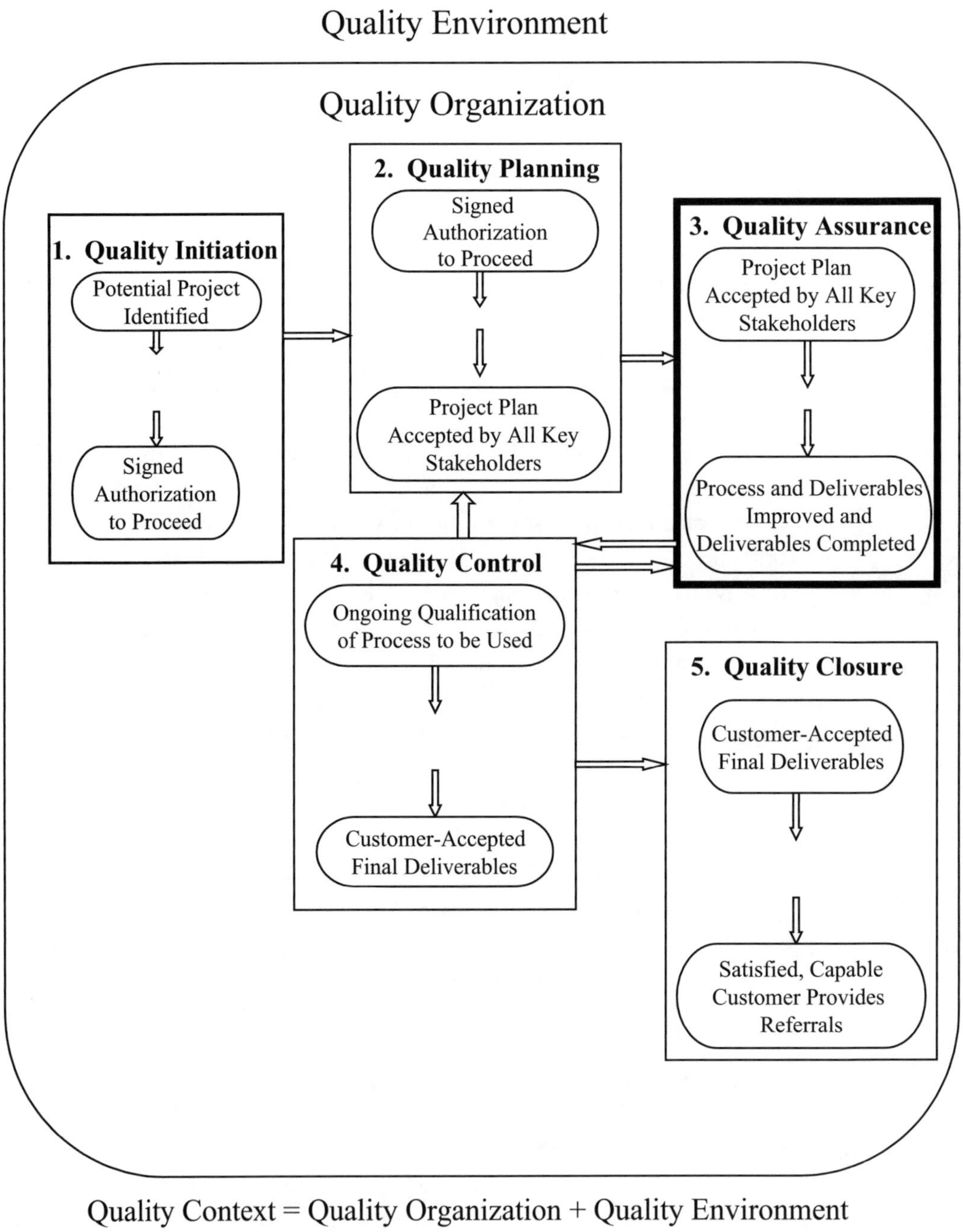

Quality Environment

Quality Organization

2. Quality Planning
- Signed Authorization to Proceed
- Project Plan Accepted by All Key Stakeholders

1. Quality Initiation
- Potential Project Identified
- Signed Authorization to Proceed

3. Quality Assurance
- Project Plan Accepted by All Key Stakeholders
- Process and Deliverables Improved and Deliverables Completed

4. Quality Control
- Ongoing Qualification of Process to be Used
- Customer-Accepted Final Deliverables

5. Quality Closure
- Customer-Accepted Final Deliverables
- Satisfied, Capable Customer Provides Referrals

Quality Context = Quality Organization + Quality Environment

FIGURE 4-2 Project Quality Assurance Flowchart

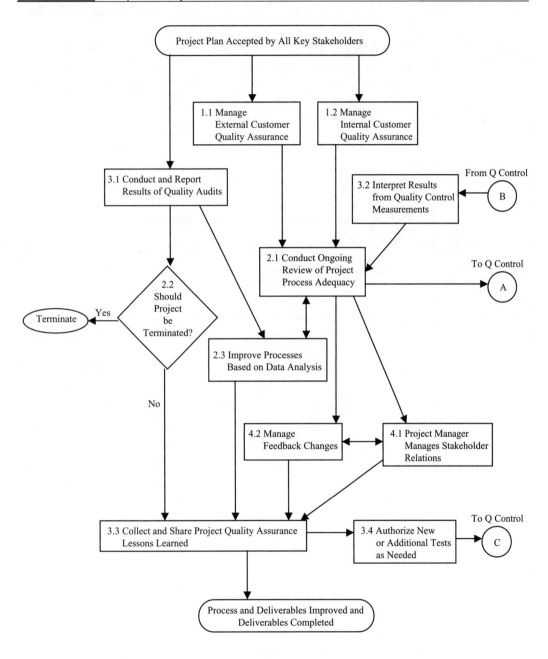

TABLE 4-1 Project Quality Assurance Factors Table

Pillar	Activities	Tools
1. Customer Satisfaction	1.1 Manage External Customer Quality Assurance 1.2 Manage Internal Customer Quality Assurance	Customer Significance-Success Matrix
2. Process Improvement	2.1 Conduct Ongoing Review of Project Process Adequacy 2.2 Conduct Interim Project Termination Review 2.3 Improve Processes Based on Data Analysis	Process Qualification Levels Interim Project Termination Review Project Check Sheet, Project Histogram
3. Fact-Based Management	3.1 Conduct and Report Results of Quality Audits 3.2 Interpret Results of Quality Control Measurements 3.3 Collect and Share Project Quality Assurance Lessons Learned 3.4 Authorize New or Additional Tests as Needed	Project Quality Audit Control Chart Plus Delta Model
4. Empowered Performance	4.1 Project Manager Manages Stakeholder Relations 4.2 Manage Feedback Changes	Change Control Form

FIRST PROJECT QUALITY PILLAR: CUSTOMER SATISFACTION

Project quality assurance includes improving the management of external and internal customer satisfaction expectations. These are ongoing activities that occur throughout the entire project quality assurance stage of the project. Improving the satisfaction of external customers contributes to bottom-line profitability by generating repeat business and positive referrals. Improving the satisfaction of internal customers increases operational efficiency, accelerates the pace of organizational learning, and supports meaningful teamwork. Properly managing the human resource system sustains intrinsic work motivation and cooperative empowered performance of individuals and teams dedicated to quality assurance.

1.1 Manage External Customer Quality Assurance

Managing *external customer quality assurance* involves continually engaging in the following project activities:

1. Defining and segmenting customers and markets
2. Listening to and learning from customers
3. Linking customer input to design, production, and delivery processes

4. Building trustworthy relationships from initial contact to follow-up services

5. Collecting and responding to customer complaints systematically

6. Measuring perceptions of quality with benchmarking techniques and improving service accordingly.

Managing external customer quality assurance requires that project customer information be acquired by any or all of the following methods: comment cards and formal surveys, focus groups, field intelligence, direct customer contact, complaints analysis, and Internet monitoring.

Managing external customer quality assurance ultimately means providing the product/service during moments of truth so that the perception of quality is solidified in the customer's mind. For example, when a customer purchases a Lexus automobile from a dealership, every contact from sales to service affects the perceived quality of that product and the company. In addition, a good project team builds project customer relationships by providing customers with: easy accessibility; strong commitments: well-trained, empowered customer-contact employees; and rapid, effective complaint response time.

Next, external customer quality assurance requires supplier certification and mutual process disclosures for continued improvement of raw materials and other project inputs. The "garbage in, garbage out" syndrome is overcome by disallowing any substandard upstream supplier input so that external downstream customer satisfaction will be better assured.

Finally, measuring customer satisfaction by means of the *customer significance-success (CSS) matrix* guides improvement efforts (see Figure 4-3). Strength in managing external customer assurance means demonstrating successful delivery of products/services in a way that meets or exceeds customer expectations on issues that are of high significance. If, however, resources are wasted on successful delivery of insignificant features or significant customer expectations are neglected, the organization is ultimately vulnerable to competitors who can manage external customer assurance more effectively.

1.2 Manage Internal Customer Quality Assurance

Managing *internal customer quality assurance* involves human resource practices that inspire confidence in the project system processes. The leading human resource practices in project quality assurance include:

1. Integrating human resource development plans into organizational and project objectives

FIGURE 4-3 Customer Significance-Success Matrix

Customer Significance	Provider Success	
	Low	High
High	Vulnerable	Strength
Low	Does Not Matter	Overkill

2. Designing project work to promote personal and organizational learning, innovation, and flexibility
3. Implementing project performance management subsystems that recognize and reward excellence
4. Promoting cooperative teamwork and individual empowerment to ensure project customer satisfaction
5. Investing in human resource training, education, and well-being to support project productivity
6. Listening to and measuring the voice of the employee, and improving human resource satisfaction indices accordingly.

With regard to designing project work, meaningfulness is enhanced by skill variety, task identity, and task significance. Productivity is enhanced by increased autonomy and performance feedback. Project managers who expand opportunities for team involvement through suggestion subsystems and other improvement activities are likely to increase internal customer project assurance levels.

Furthermore, performance appraisal systems that factor in both system-determined and individual contributions to performance are important. The 360-degree feedback process allows for peer review, subordinate input, customer evaluations, self-assessments, and personal development plans so that employees are more likely to regard the appraisal process as fair and developmental.

SECOND PROJECT QUALITY PILLAR: PROCESS IMPROVEMENT

While process improvement in previous stages consisted mostly of planning, now process improvement becomes more action-oriented. The project quality assurance stage requires ongoing review of the adequacy of existing processes, an interim project termination review, and ongoing improvement of processes based on data analysis.

2.1 Conduct Ongoing Review of Project Process Adequacy

The process improvement practices that reassure key project stakeholders include:

1. Clearly translating project customer requirements into project design
2. Using appropriate quality tools to implement incremental, competitive-parity, or breakthrough improvements
3. Ensuring that supplier requirements are met and new partnering relationships are formed to increase project efficiency
4. Identifying statistically significant variations in project performance
5. Accurately analyzing the root causes of variations, making corrections, and verifying new project operation results
6. Measuring and benchmarking project processes for continual improvement.

The project process qualification activities are also ongoing to determine whether a particular process is at Spontaneous Level 1, Initialized Level 2, Formalized Level 3, or Optimized Level 4 (see Figure 1-4).

The aim of project process assurance is to improve qualification levels as soon as possible and confirm that qualified processes are in fact being implemented on a regular basis. This confirmation goes a long way toward one of the primary aims of quality assurance—to provide confidence that the project will satisfy relevant quality standards.

2.2 Conduct Interim Project Termination Review

One of the assurance decisions that is crucial is the *interim project termination review*. If a preponderance of midstream data indicate that the project is not worth further investment of resources, a rigorous termination review will result in a recommendation to cut losses and reallocate resources to more promising options.

Among the considerations that might result in interim termination are the following:

- Low probability of achieving technical objectives and commercializing outcomes
- Technical and/or production problems that cannot be solved with available resources
- Unanticipated large cost overruns or reduced profitability
- Unacceptable schedule delays
- Lowered market potential due to substitutes and competitors
- Shifts in strategic priorities
- Top management decisions to outsource or subcontract the project to cost cuts
- Problems in protecting new project knowledge through patents
- Emergence of a better alternative for use of funds and resources.

Project managers need to be aware of these threats to interim project termination and be prepared to respond to each in an appropriate manner. While many of these factors do not suggest poor management of the project, they do suggest that terminating the project is in the best interests of the overall organization. When confronted with this truth, project managers and sponsors must remember that while they are advocates for the project, they must make recommendations and decisions that are correct for the entire organization. Terminating a project is often painful, yet sometimes necessary.

2.3 Improve Processes Based on Data Analysis

Once the project has passed the interim review, new process resources need to be allocated to continually improve the efficiency of the endorsed project through data collection and analysis. Among the quality tools to be used are project check sheets and project histograms.

Project check sheets are special types of data collection forms in which analytical results may be easily interpreted on the form without additional processing. The sample of a project check sheet in Figure 4-4 depicts the frequency of different types of problems on a weekly basis and helps to quickly identify the most frequently occurring problem as problem C.

In addition to project check sheets to improve processes based on data analysis, the *project histogram* is a quality tool that graphically depicts the frequency or number of observations of a particular value that occurs within a specific group. Figure 4-5 provides a graphic example of how frequently a particular range of thickness occurred. The greatest frequency of 30 occurrences is visually shown in the population of objects with a thickness of at

FIGURE 4-4   Project Check Sheet

Problem	Week			
	1	2	3	Total
A	I I	I I	I	5
B	I	I	I	3
C	I I I I	I I	I I I I	10
Total	7	5	6	18

least 9.4 but less than 9.6. Using project histograms to improve processes, however, requires that the tool be used under typical conditions and consist of a minimum sample size of 50.

THIRD PROJECT QUALITY PILLAR: FACT-BASED MANAGEMENT

Among the fact-based management tasks of project quality assurance are (1) conducting and reporting quality audits, (2) interpreting quality control measurements, (3) collecting quality assurance lessons learned, and (4) authorizing new or additional tests as needed.

FIGURE 4-5   Project Histogram

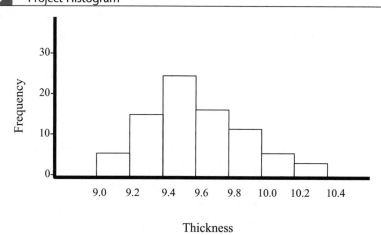

Thickness

3.1 Conduct and Report Results of Quality Audits

Internal or external *project quality audits* focus on identifying whether documented processes are being followed and are effective, and reporting unacceptable variances to project managers for correction. Audits normally include a review of project process records, training records, registered complaints, documented suggestions, corrective actions, and issues from previous audit reports.

The audit normally begins by asking those who perform a project process regularly to explain how it works. Their statements are compared to written procedures and project norms, and compliance and deviations are noted. Next, the paper trail and other data streams are followed to determine whether the project process is consistent with the intent of the written procedure and the operator's explanation.

Quality audits also go beyond routine procedures of quality control and address the following strategic questions:[2]

1. To what extent are current quality policies and goals aligned with organizational mission priorities?
2. To what extent does the current level of quality provide product/ service satisfaction to customers?
3. To what extent is the current level of quality externally competitive with the moving target of the marketplace?
4. To what extent is the organization making progress in reducing and/or eliminating the costs of poor quality?
5. To what extent is the cross-functional collaboration among and between functional departments adequate to ensure optimal business results?
6. To what extent does the current level of quality performance meet social and environmental sustainability standards?

3.2 Interpret Results of Quality Control Measurements

One of the important tasks of fact-based management during the project quality assurance stage is to rigorously apply statistical thinking in the interpretation of quality control measurements. The temptations during this stage are either to overreact or to underreact in an effort to reassure customers that quality service is being provided.

To counteract these temptations, project managers need to interpret control chart data in a way that focuses attention on statistical outliers above the upper control limits and below the lower control limits as objects of intervention. Project managers who are not trained in statistics will often overreact to productivity differences in project team member performance that are not statistically significant, and thus create problems where there were none. In addition, project managers not trained in statistics may ignore substandard performance below the lower control limit and permit a statistically important productivity problem to persist.

Data and trends are only useful if they are properly interpreted statistically so that accurate information or facts provide the solid foundations for improvement decision-making.

3.3 Collect and Share Project Quality Assurance Stage Lessons Learned

Just as lessons at the end of the initiation and planning stages were collected, the same procedure can and should be followed near the end of the project quality assurance stage. Lessons are to be used to improve future stages of the current projects, and, through sharing, contribute to the organization's learning capacity.

3.4 Authorize New or Additional Tests As Needed

During the quality assurance stage, project managers not only need to interpret data from existing tests, but also must be prepared to authorize new or additional tests as needed. As customers request midstream changes in product/service design, new or additional tests may be necessary to assure the process and outcome quality of the contracted product/service. For example, if a Dell Computer customer changes her mind and wants more "bells and whistles" than her initial order, the project quality assurance team will need to perform new and additional tests to ensure that those new "bells and whistles" work properly.

FOURTH PROJECT QUALITY PILLAR: EMPOWERED PERFORMANCE

In the quality assurance stage, empowered performance is enhanced through the management of positive stakeholder relationships as well as the management of feedback changes.

4.1 Project Manager Manages Stakeholder Relations

Project managers empower key project stakeholders in a variety of ways; two important ways are sharing information and eliciting discretionary effort. Although key project stakeholders have clarified their role responsibilities endorsed the project, an ongoing, accurate stream of information that keeps these stakeholders apprised of interim progress is important. The project manager is able to elicit discretionary effort from the key project stakeholders by soliciting midstream feedback and sharing it with project participants to help motivate them.

FIGURE 4-6   Project Change Request Form

Part One (to be filled out by Submitter)

Submitter: _____ Phone _____ Email _____
Date of Submission: _____
Type of Change: Problem _____ Enhancement _____
Recommended Priority: Critical _____ High _____ Medium _____ Low _____
Description of proposed change:

Reason for change:

Estimated schedule impact:
Estimated cost impact:

Part Two (to be filled out by Technical Leads)

Technical Area	Estimated Impact	Recommend (yes or no)	Signed
A	_____	_____	_____
B	_____	_____	_____
. . .	_____	_____	_____
X	_____	_____	_____

Part Three (to be filled out by Project Manager)

Approved? _____   date _____
Rejected? _____   why? _____
Deferred? _____   what more info needed? _____
Assigned priority: Critical _____ High _____ Medium _____ Low _____
Responsibility assigned to: _____
Tracking number: _____
Date completed: _____

Project Quality Control

Quality control is defined in the *PMBOK® Guide* as "monitoring specific project results to determine if they comply with relevant standards and identifying ways to eliminate causes of unsatisfactory performance."[1] Control is the activity of ensuring conformance to standards and taking corrective action when necessary to correct problems. Long-term improvements to a process cannot be made until the process is first brought under control.

We define quality control activities to start when processes are qualified in quality assurance. This is an ongoing activity, so quality control activities start repeatedly during a typical project. Quality control activities should continue until the customer accepts the final project deliverables. Quality control is the fourth stage in the five-stage project quality process model, as shown in Figure 5-1.

The project quality control and project quality assurance stages have a large degree of concurrent interaction. For example, if test results are excellent in the project quality control stage, the project deliverables could be accepted rapidly and the project would move to the quality closure stage. On the other hand, if test results are not excellent, more process work in the quality assurance stage may be necessary or some replanning may need to occur back in the quality planning stage.

As is true for all the process stages, the level of detail needed during the project quality control stage can vary significantly from one project to another. Typical project quality control activities are shown in Figure 5-2. The flow of information to the project quality planning stage is shown. The flows of information both to and from the project quality assurance stage are also depicted in this flowchart.

Many tools and checklists can be used to accomplish the tasks during this process stage. Table 5-1 shows the project quality pillars, activities, and tools to be used during the project quality control stage.

This chapter is structured to follow the order of project quality pillars and their sequenced activities shown in Table 5-1. The first number of the

FIGURE 5-1 Project Quality Process Model

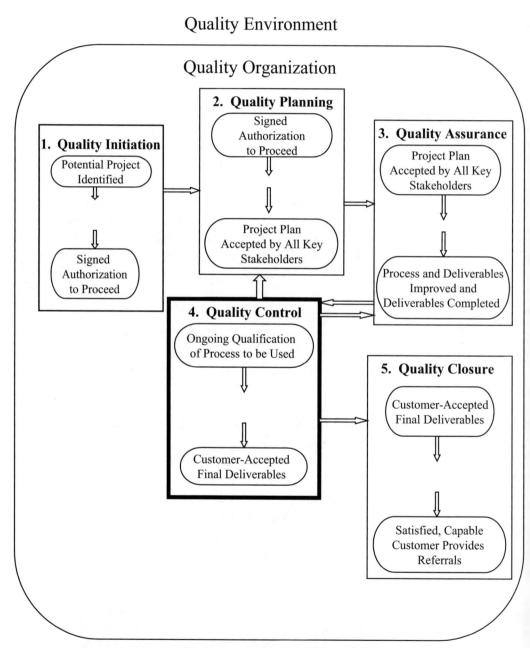

Quality Context = Quality Organization + Quality Environment

FIGURE 5-2 Project Quality Control Flowchart

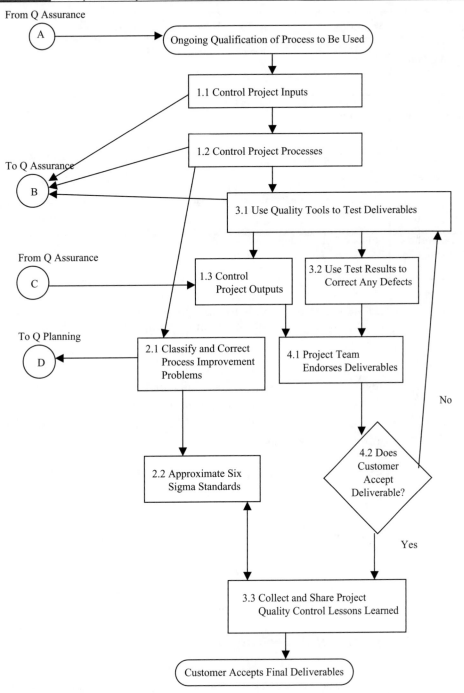

TABLE 5-1 Project Quality Control Factors Table

Pillar	Activities	Tools
1. Customer Satisfaction	1.1 Control Project Inputs 1.2 Control Project Processes 1.3 Control Project Outputs	Control Charts
2. Process Improvement	2.1 Classify and Correct Process Improvement Problems 2.2 Approximate Six Sigma Standards	
3. Fact-Based Management	3.1 Use Quality Tools to Test Deliverables 3.2 Use Test Results to Correct Any Defects 3.3 Collect and Share Project Quality Control Lessons Learned	Flowcharts, Run Charts, Pareto Diagrams Plus Delta Model
4. Empowered Performance	4.1 Project Team Endorses Deliverables 4.2 Customer Accepts Deliverables	

listed activities corresponds to the appropriate quality pillar, e.g., Activity 1.1 is associated with the first pillar and Activity 2.1 is associated with the second pillar. The second number of the listed activity refers to the typical approximate chronological sequence of its execution within the pillar's domain, although this sequential order may vary with different projects, organizations, or industries. For example, 3.1 Use Quality Tools to Test Deliverables normally comes before 3.3 Collect and Share Project Quality Control Lessons Learned.

FIRST PROJECT QUALITY PILLAR: CUSTOMER SATISFACTION

The range of quality control includes both product results, such as deliverables, and project management results, such as budgeted cost and schedule deadlines. To ensure customer satisfaction, the project quality control stage entails prevention, inspection, and testing at three points:

1. At the receipt of incoming project resources
2. During the project creating/delivery process
3. Upon completion of project production.

1.1 Control Project Inputs

To ensure customer satisfaction and adherence to the project quality policy, several steps need to be taken. These steps include implementing procedures for project specification and design control, preventing purchasing errors, and applying inspection and sampling techniques. If incoming project resources (human and non-human) are of poor quality, the quality of the final product and the project outcome will be in jeopardy. Clarifying project specifications and design control standards is crucial for controlling the project input structure. Preventing project purchasing errors is important since proper evaluation and selection of suppliers means that initial inputs are likely to conform to requirements.

In addition, inspection and sampling techniques ensure that substandard input does not enter the processing stage. Spot-check procedures select a fixed percentage of a lot for inspection prior to acceptance determination. One hundred percent inspection is theoretically possible but usually impractical. Acceptance sampling takes a statistically calculated random portion, and uses a decision to determine lot acceptance or rejection based on the number of nonconforming items.

1.2 Control Project Processes

Since unwanted variation can occur during the process of product manufacturing or service delivery in a project, controlling in-process attributes and variables is important. An *attribute* is a performance characteristic that is either present or absent, e.g., the number of tasks completed by the project team or the number of customer complaints about the project. A *variable* is a performance characteristic that occurs in degrees, e.g., degree of conformance to standards by means of averages or standard deviations, customer time waiting for project completion, or time for a project team to move from conception to completion.

This is where it becomes essential to set *tolerances* (the result is acceptable to customers if it falls within the range specified by the tolerance) and *control limits* (the process is in control if the result falls within the control limits) properly. All project processes will demonstrate some variation. Therefore, it is important to determine what type of variation is occurring. Recall from Chapter 1 that random or normal variation occurs when many small things happen. Looking at the entire process systematically reduces random variation.

The other type of variation is called assignable cause or special variation. It occurs when a particular event that is unusual happens. Quickly identifying the unusual occurrence and making a change in the process so it cannot happen again is the best way to reduce this assignable cause variation.

The proper use of quality control charts will assist the manager in distinguishing between random and assignable causes of variation. Control charts can be used to monitor:

- Project cost variances
- Project schedule variances
- Volume and frequency of scope changes
- Errors in project documents
- Other process results.

1.3 Control Project Outputs

Final inspection of product and project outputs consists of (1) testing, and (2) taking corrective action based upon the test results. If, for example, a project product is the correct assembly of a computer, one functional test is to turn the computer on and make sure it operates properly.

Another obvious test is visual inspection. Visual inspection can be enhanced by:

- Limiting the number of quality inspection factors to five or six
- Minimizing distracting influences or time pressures
- Providing detailed instructions and checklists for inspection tasks
- Providing suitable working conditions for the final inspection activity.

Taking necessary corrective action in a timely fashion prior to delivering the product/service to the customer is expected. Any rework based on data regarding feedback changes, process improvements, and/or minor adjustments all impact customer satisfaction.

SECOND PROJECT QUALITY PILLAR: PROCESS IMPROVEMENT

In the project quality control stage, process improvement begins with the final, careful classification of process quality problems and the application of Six Sigma approaches.

2.1 Classify and Correct Process Quality Problems

Project quality process problems can normally be classified as structured, semistructured, and unstructured with respect to the volume and accuracy of information about them. Process improvement of structured problems requires that simple adjustments to the size or volume of an object can be addressed in a straightforward manner to ensure conformance with specifications. In other words, the original system of structured processes was fine but deviations needed to be corrected so that the system could be restored to its intended mode of functioning.

Semistructured and unstructured project process problems, however, require more creativity. The four general types of these problems are:[2]

1. Unstructured performance problems
2. Efficiency problems
3. Project product design problems
4. Project process design problems.

An example of an unstructured performance problem is that sales to project customers may be lagging, but there is no one standard way of selling. It is difficult to correct selling performance by referral to a nonexistent standard. Sophisticated diagnosis and creative solutions are required.

An example of an efficiency process problem is slow and cumbersome development of a needed part. Internal organizational stakeholders are adversely impacted by project processes and require coordinated operational streamlining solutions.

In the third type of process problem—project product design—the design does not mesh with customer expectations. A more detailed application of the house of quality approach may be warranted.

Finally, in the fourth type of problem—project process design—radical redesign of the process may be necessary to be competitive. Benchmarking and reengineering can be used to address this process improvement challenge.

2.2 Approximate Six Sigma Standards

Managers of project processes should strive for excellence. The most demanding current method of striving for process excellence is called Six Sigma. *Six Sigma* represents a quality process level of at most 3.4 defects per million opportunities (3.4 dpmo). While many projects would not have anywhere near this volume of defect opportunities, striving toward this level of process capability can and should be the aim of project process improvement during the project quality control stage.

To move toward Six Sigma level quality, every opportunity for a failure to meet customer expectations needs to be identified and tracked. The aim is to analyze its root cause, take corrective action, and reduce the defects as much as possible. This approach also entails extensive statistical process control and Six Sigma training to create in-house experts who can lead teams and apply the appropriate tools/metrics that focus on bottom line business results. Once critical-to-quality elements are identified in the final process improvement, they are analyzed, improved, and brought under control as rapidly as possible.

THIRD PROJECT QUALITY PILLAR: FACT-BASED MANAGEMENT

Project managers should ensure that several fact-based management activities occur during project quality control. These activities include ensuring that quality tools are used to test deliverables, that test results are used to correct any final defects, and that project quality control lessons are shared to enhance future organizational learning.

3.1 Use Quality Tools to Test Deliverables

Project team members can use a wide range of quality control tools to test the final deliverables. Some representative tools include: elementary and advanced statistical tools, quality function deployment tools, process capability tools, process managing, mistake proofing, and organizational quality audits.

Flowcharts, run charts, and Pareto diagrams are three examples of the standard quality control tools that are useful in implementing fact-based management. The flowcharting illustrated throughout this book demonstrates the way that process managing can clarify the "operational mess" that often constitutes quality control efforts. Flowcharting identifies the sequence of activities and their interconnections from start to finish. By looking at a flowchart, all key project stakeholders can graphically grasp "the big picture" and their role in the project's completion. In addition, flowcharting enables a project team to pinpoint obvious problems, error-proof the processes, streamline the project by eliminating non-value-added steps, and focus variation reduction interventions.

Project run charts are line graphs in which data are plotted over time. The vertical axis represents a measurement; the horizontal axis is the time dimension. Run charts depict the project performance and the variation of the process indicator over time.

Figure 5-3 depicts the run chart of the weekly percentage of on-time task completions of a project team. The run chart indicates a progressively higher percentage of project team on-time task completions while depicting a fluctuating variation from week to week. Further analysis of these fluctuations might well provide opportunities for improving project team performance by controlling for unnecessary fluctuations.

FIGURE 5-3   Project Run Chart

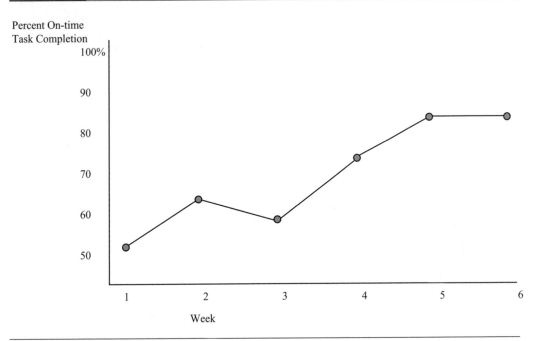

Percent On-time
Task Completion

Finally, project *Pareto diagrams* are used to depict data collected on check sheets. The characteristics observed are ordered from largest frequency to smallest. For example, Figure 5-4 illustrates the relative magnitude of each type of defect. This knowledge can quickly be used to prioritize control and improvement efforts. The stratification of data can also lead to progressive understanding and resolution of semistructured and unstructured control problems.

3.2 Use Test Results to Correct Any Defects

The technical testing and results are used to correct any defects. The number of non-conforming project outputs may warrant extensive rework or immediate project/process adjustments. *Rework* is action taken to bring a defective or nonconforming item into compliance with specifications. Since this is a major cause of project overruns, the project team should make every reasonable effort to minimize rework. Nevertheless, fact-based management of quality control is not completed until inspected items are accepted.

In addition to these selected quality control tools, a wide variety of commercial software is available that implements the full range of statistical

FIGURE 5-4   Project Pareto Diagram

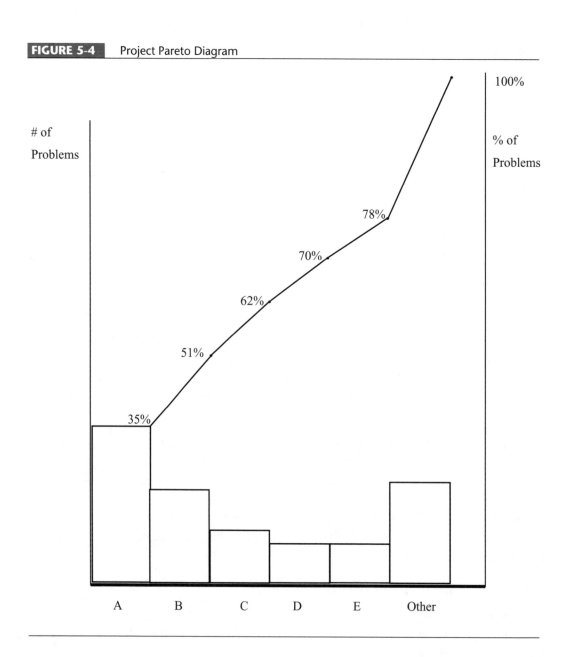

process control tools. Annual software surveys can be found in such professional publications as *Quality Progress* and *Quality Digest*. Commercial software to aid in a variety project control tasks can be found in periodic surveys published in *PM Network*.

3.3 Collect and Share Project Quality Control Lessons Learned

The collection and sharing of project quality control lessons can be facilitated through use of the plus delta tool. The lessons learned can be shared to promote organizational learning and improve future project control activities.

FOURTH PROJECT QUALITY PILLAR: EMPOWERED PEFORMANCE

The two major activities that empower performance during project quality control both deal with the project deliverables. First, the project team must endorse the deliverables and then the customer must accept the deliverables.

4.1 Project Team Endorses Deliverables

An important dimension of project team empowerment is individual and collective accountability for the quality of the final deliverables. One of the surest ways to enhance team empowerment is through justified shared pride in (1) the project processes used to produce the project deliverables, and (2) the quality of the final deliverables themselves. Successful project leaders and their project teams develop a history together and leave a legacy in that they will only hand over to the customer acceptable final deliverables. This acceptability is determined by the project team endorsing the quality of the deliverables prior to asking the customer to accept them. Nurturing this sense of collective honor creates a culture committed to project quality in the future.

4.2 Customer Accepts Deliverables

After the project team endorses the final deliverables, the customer is handed the project deliverables. However, the project quality control stage is not complete until the customer accepts the deliverables. If the customer has reservations about acceptance, these need to be addressed until the customer is satisfied. Prudent project managers and customers will often require written acceptance of the deliverables.

Project quality assurance is complete upon customer acceptance of the final deliverables. The project, however, is not complete. The final stage of project quality closure is important, yet frequently short-changed since many project participants are already working on new projects.

NOTES

1. Project Management Institute Standards Committee, *A Guide to the Project Management Body of Knowledge* (*PMBOK® Guide*) (Upper Darby, PA: Project Management Institute, 2000), p. 95.
2. James R. Evans and William M. Lindsay, *The Management and Control of Quality*, 5th edition (Cincinnati, OH: South-Western Publishing, 2002).

CHAPTER 6

Project Quality Closure

C losure is defined in the *PMBOK® Guide* as "formalizing acceptance of the project and bringing it to an orderly end."[1] The quality closure stage begins with the customer's formal acceptance of the final project deliverables and ends with referrals from the capable, satisfied customer. Quality closure follows quality control as the final stage in the five-stage project quality process model shown in Figure 6-1.

As is true of all five stages, the management activities will be much more involved on some projects than on others. Large, complex, unfamiliar projects will require more in-depth closure procedures than smaller, simpler, more familiar projects. The typical quality closure activities required are depicted in the flowchart in Figure 6-2.

Several factors facilitate the movement from customer acceptance of final project deliverables to the point at which the now-satisfied customer provides referrals. These include the role of the project quality pillars in closure, completion of necessary activities, and correct use of project quality tools. Table 6-1 categorizes these pillars, activities, and tools for the project quality closure stage into the project quality closure factors table.

This chapter is structured to follow the order of project quality pillars and their sequenced activities shown in Table 6-1. The first number of the listed activities corresponds to the appropriate quality pillar, e.g., Activity 1.1 is associated with the first pillar and Activity 2.1 is associated with the second pillar. The second number refers to the typical approximate chronological sequence of its execution within the pillar's domain, although this sequential order may well vary with different projects, organizations, or industries. For example, 3.1 Assess Overall Project Results normally comes before 3.3 Collect, Share, and Document Overall Project Lessons Learned.

FIRST PROJECT QUALITY PILLAR: CUSTOMER SATISFACTION

Throughout project closure, all key project participants must be kept aware of the customer expectations initially mentioned during project initia-

FIGURE 6-1   Five-Stage Project Quality Process Model

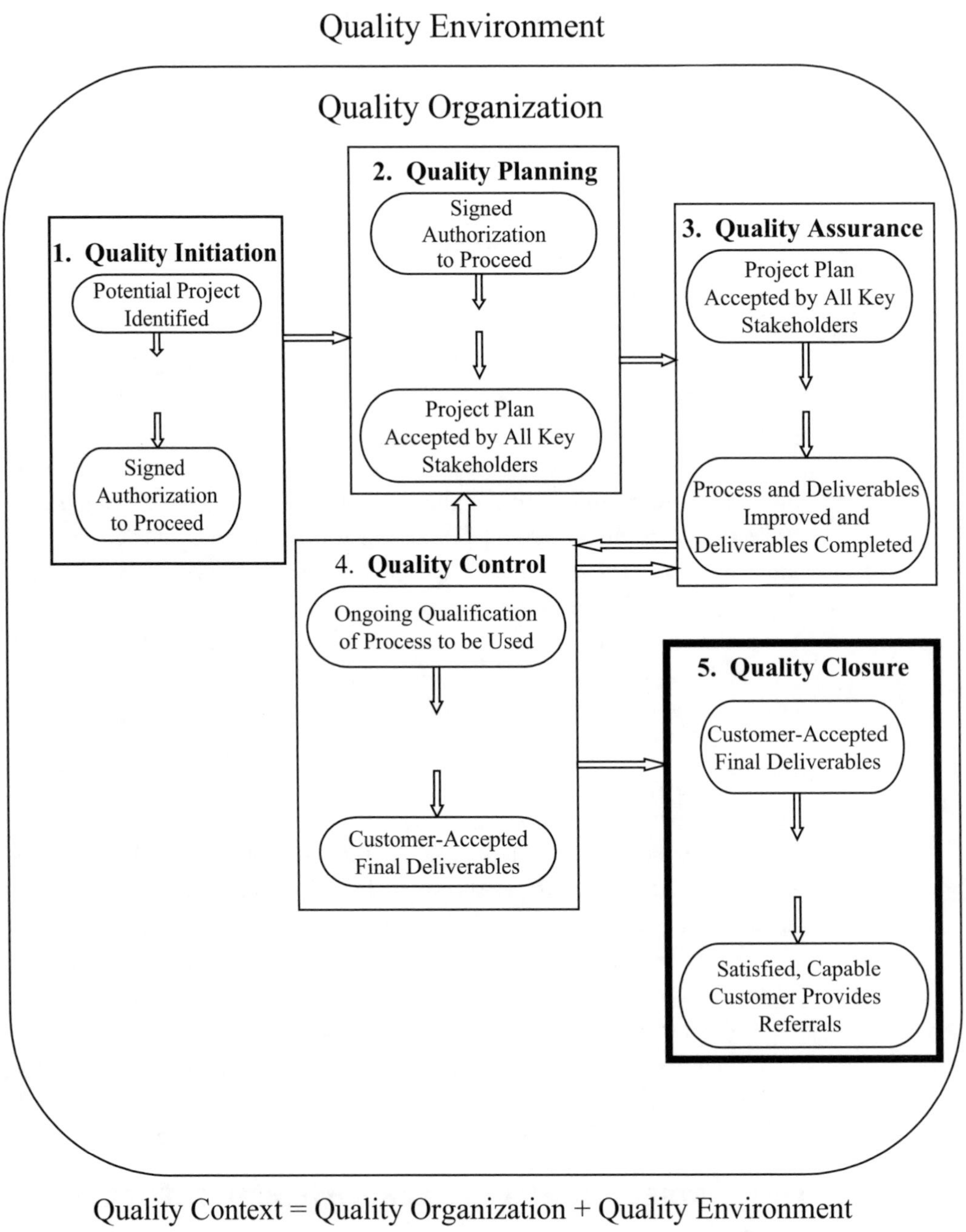

Quality Environment

Quality Context = Quality Organization + Quality Environment

FIGURE 6-2   Project Quality Closure Flowchart

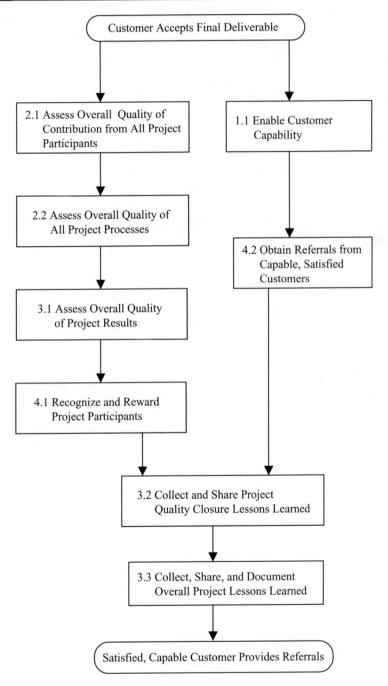

TABLE 6-1 Project Quality Closure Factors Table

Factor	Activities	Tools
1. **Customer Satisfaction**	1.1 Enable Customer Capability	
2. **Process Improvement**	2.1 Assess Overall Quality of Contributions from All Project Participants 2.2 Assess Overall Quality of All Project Processes	Process Qualification Levels
3. **Fact-Based Management**	3.1 Assess Overall Project Results 3.2 Collect and Share Project Quality Closure Lessons Learned 3.3 Collect, Share, and Document Overall Project Lessons Learned	Plus Delta Model Plus Delta Model
4. **Empowered Performance**	4.1 Recognize and Reward Project Participants 4.2 Obtain Referrals from Capable, Satisfied Customers	

tion and further refined throughout the project. Regardless of pressures concerning any of the other five objectives (cost, schedule, scope, contribution to the organization, and contribution to society), quality should always be at least a co-equal objective during project closure. The very reason for undertaking a project is to create a product or service that is useful to the customer. It would be fruitless to forget that now.

Thus, the most important customer satisfaction activity during the project quality closure stage is enabling customer capability. Turning over a quality deliverable to a customer who is not fully capable of using it is not a quality ending to a project.

1.1 Enable Customer Capability

Growing dimensions of project customer satisfaction are support service and training opportunities after delivery of project deliverables. Successful computer manufacturers, for example, realize that support services and training available after the sale may determine whether the sale takes place at all. This support and training affect the level of computer customer satisfaction.

Project quality closure requires the two customer Cs: capability and co-partnership. *Customer capability* means making an assessment of the customer's readiness to use the project deliverable properly. This assessment should be made as soon as practical so additional support services and training can be made available to the customer. In addition, providing support services and training helps develop a copartnering relationship between the project team and the customer. The customer realizes that the project team has a stake not only in a particular deliverable, but also in the sustained coprosperity of the customer who uses the deliverable. In turn, the enabled capable customer becomes a copartner in expanding the positive referral network for future project work.

SECOND PROJECT QUALITY PILLAR: PROCESS IMPROVEMENT

During the project quality closure stage, two process improvement activities must occur: (1) assessment of the overall quality of contributions from all project participants, and (2) assessment of the overall quality of all project processes.

2.1 Assess Overall Quality of Contributions from All Project Participants

The evaluation of the quality of contributions from all project participants provides appreciative feedback for past performance and for future improvement. Part of project closure assessment is the private and public expression of appreciation for contributions made. Appreciated project participants can be enthusiastic participants on future projects. Unappreciated project participants soon become nonparticipants.

Furthermore, critical feedback can enhance project team contributions in future projects and adds to the team member's professional development. This evaluation also helps the project manager in selecting members for future projects and matching them with role responsibilities that will leverage their talents and stretch their capabilities.

2.2 Assess Overall Quality of All Project Processes

In addition to evaluating participant contributions, part of project closure process improvement is a final assessment of the overall quality of all project processes. Some of the project processes probably were excellent, while others probably were not. Some might have been eliminated to reduce cycle time and save resources. Others might have been marginally acceptable but could have been improved. If some had been improved earlier in the project, perhaps schedule time and costs could have been reduced. Still other processes might have been drastically improved had they been computerized.

THIRD PROJECT QUALITY PILLAR: FACT-BASED MANAGEMENT

The fact-based management portion of the project quality closure stage is threefold: (1) an assessment of overall project results, (2) the collection and sharing of project quality closure lessons learned, and (3) the collection, sharing, and documentation of overall project lessons learned.

3.1 Assess Overall Project Results

Evaluating overall project results is normally based upon the impact that is realized and predicted at this time. However, the full impacts of the project to the organization and society are rarely realized at the time of customer acceptance of final deliverables. Often a distinction can be made between verified short-term results and estimated long-term results.

Using the metrics established at the beginning of the project, the team will be able to clearly document results in both project and organizational outcomes. Positive project outcomes may include reduced costs, schedule time, and complaints. Positive organizational outcomes may include monetary contributions to the financial bottom line, elimination of organizational inefficiencies, quicker response time, and other documented gains from benchmarking.

3.2 Collect and Share Project Quality Closure Lessons Learned

The collection and sharing of project quality closure lessons can be facilitated through the use of the plus delta tool. The stage-specific lessons can be shared to promote organizational learning and improve future project quality closure activities.

3.3 Collect, Share, and Document Overall Project Lessons Learned

While lessons learned should have been collected periodically—usually at the end of each project stage, it is also important to capture and analyze lessons learned from the perspective of the entire, completed project. Often the total is greater than the sum of the parts. Collecting, documenting, and sharing overall lessons learned is a time-consuming process, but it is usually one of the most organizationally valuable activities. Project lessons can be placed in an organizational data base and serve to enhance organizational memory and accelerate cross-functional organizational learning.

FOURTH PROJECT QUALITY PILLAR: EMPOWERED PERFORMANCE

Recognition and reward celebrations for project participants at the end of the project are activities that empower performance both for the current project (people may work harder in anticipation) and for future projects. Obtaining referrals from capable, satisfied customers is also very empowering since it means there is "life after the project" for many of the participants.

4.1 Recognize and Reward Project Participants

Recognition celebrations and reward ceremonies are appropriate vehicles for reinforcing good project performance and empowering future performance. Most project teams and organizations "under celebrate" project successes. This is a closure mistake. Excellent performers need and deserve recognition and rewards. Enhancing their self-respect and self-esteem is key in motivating them to engage in future projects.[2]

Organizations that are driven from one project to the next without celebrating their project successes soon burn people out. They disempower themselves by neglecting to celebrate successes.

4.2 Obtain Referrals from Capable, Satisfied Customers

Successful project managers who last for many years obtain referrals from capable and satisfied customers. They indicate their availability for return engagements and usually have more repeat business than others. Project managers who neglect this project quality closure responsibility have fewer opportunities for future leadership.

World-class project managers value referrals and repeat engagements. These managers and their teams enhance their reputations and create strategic competitive advantages for consideration on future projects.

NOTES

1. Project Management Institute Standards Committee, *A Guide to the Project Management Body of Knowledge* (*PMBOK® Guide*) (Upper Darby, PA: Project Management Institute, 2000), p. 30.
2. Morris Altman, *Worker Satisfaction and Economic Performance* (Armonk, NY: M.E. Sharpe, 2001).

CHAPTER 7

Summary and Challenges

P*roject quality management* truly is the merging of the two fields of project management and quality management. It is more than a knowledge area in project management and more than a means of better planning and managing improvement projects in quality. It is the systematic adaptation and use of quality tools and knowledge to meet the unique needs of projects.

The four project quality pillars of customer satisfaction, process improvement, fact-based management, and empowered performance are useful for structuring the activities and tools treated in each stage of the five-stage project quality process model: project quality initiation, project quality planning, project quality assurance, project quality control, and project quality closure. The structure of this dual field integration is summarized in the *integrated project quality activity matrix* presented as Figure 7-1.

While all of the activities shown in Figure 7-1 should be performed at some level (they can be streamlined on easy projects and may be very involved on difficult projects), some pose core project quality management challenges. The challenges arise either because these are unfamiliar activities that are not performed often in many organizations, because they are only partly performed, or because they are difficult to accomplish in and of themselves. In any event, many people do not realize the significance of each of these activities.

Each of these activities is important because the output is essential to project success. Some are also important because other essential activities depend on them. For example, in project quality initiation, the identification and prioritization of customer expectations is a core project quality management challenge both for its own sake and because the next crucial activity (align project with organizational objectives) depends on it. If more people would realize the unique challenges posed by the core project quality management activities, more project participants would accomplish these activities and, in turn, more projects would be successful.

FIGURE 7-1 Integrated Project Quality Activity Matrix

Pillar/Stage	Project Quality Initiation	Project Quality Planning	Project Quality Assurance	Project Quality Control	Project Quality Closure
1. Customer Satisfaction	1.1 **Assign Sponsor** 1.2 Select Project Manager 1.3 **Identify and Prioritize Customer Expectations** 1.4 Align Project with Organizational Objectives 1.5 Select Core Team Members 1.6 Determine Team Operating Principles	1.1 **Determine Customer Satisfaction Standards** 1.2 **Determine Customer Tradeoff Values** 1.3 Determine Levels of Decision-Making Authority	1.1 **Manage External Customer Quality Assurance** 1.2 **Manage Internal Customer Quality Assurance**	1.1 Control Project Inputs 1.2 **Control Project Processes** 1.3 Control Project Outputs	1.1 **Enable Customer Capability**
2. Process Improvement	2.1 Adopt or Develop Quality Policy 2.2 Flowchart the Overall Project 2.3 **Identify Assumptions and Risks** 2.4 Establish Knowledge Management Processes	2.1 Assess and Prioritize Improvement Needs 2.2 **Develop Project Quality Management Plan** 2.3 Plan Project Process and Product 2.4 Identify Needed Inputs and Suppliers 2.5 Quality All Project Processes 2.6 Replan as Needed	2.1 **Conduct Ongoing Review of Project Process Adequacy** 2.2 Conduct Interim Project Termination Review 2.3 Improve Processes Based on Data Analysis	2.1 Classify and Correct Process Improvement Problems 2.2 **Approximate Six Sigma Standards**	2.1 Assess Overall Quality of Contributions from All Project Participants 2.2 **Assess Overall Quality of All Project Processes**
3. Fact-Based Management	3.1 Agree to Make Fact-Based Decisions 3.2 **Identify Lessons Learned from the Past** 3.3 Collect and Share Project Quality Initiation Lessons Learned	3.1 Identify Data to Collect 3.2 Develop Project Communications Plan 3.3 Collect and Share Quality Planning Lessons Learned	3.1 **Conduct and Report Results of Quality Audits** 3.2 Interpret Results of Quality Control Measurements 3.3 Collect and Share Project Quality Assurance Lessons Learned 3.4 Authorize New or Additional Tests as Needed	3.1 Use Quality Tools to Test Deliverables 3.2 **Use Test Results to Correct Any Defects** 3.3 Collect and Share Project Quality Control Lessons Learned	3.1 Assess Quality of Overall Project Results 3.2 Collect and Share Project Quality Closure Lessons Learned 3.3 **Collect, Share, and Document Overall Project Lessons Learned**
4. Empowered Performance	4.1 Develop Ethical Work Culture Values 4.2 Select Project 4.3 **Formally Commit to Project**	4.1 **Core Team Commits to Project Plan** 4.2 Plan and Conduct Project Kick-Off Meeting 4.2 **All Key Stakeholders Commit to Project Plan**	4.1 Project Manager Manages Stakeholder Relations 4.2 **Manage Feedback Changes**	4.1 **Project Team Endorses Deliverables** 4.2 **Customer Accepts Deliverables**	4.1 **Recognize and Reward Project Participants** 4.2 **Obtain Referrals from Capable, Satisfied Customers**

The core project quality management challenges are shown in bold face type in Figure 7-1. The remainder of this chapter identifies and describes these core challenges by project stage.

PROJECT QUALITY INTITIATION CORE CHALLENGES

The five activities singled out as core challenges during project quality initiation are: assign sponsor, identify and prioritize customer expectations, identify risks and assumptions, identify previous lessons learned to be used on this project, and formally commit to the project.

The selection of a project sponsor is a core challenge. The sponsor is usually a high-level executive who does not have enough time to manage the project, but who has a strong vested interest in having the project reach a successful conclusion. The sponsor has several important project responsibilities, as shown in Figure 1-3, Project Lifecycle Accountability Matrix. A project will have a much better chance of success if a capable individual is assigned as a sponsor and is instructed regarding his or her project responsibilities.

The second core challenge during project quality initiation is to identify and prioritize customer expectations. The very reason for conducting a project is because some customer needs the project deliverables. Often in the eagerness to secure a contract, organizations perform this step inadequately and end up with a project that either is a poor fit or is planned poorly. While those are additional steps, performing them correctly depends on a good understanding of customer expectations.

The third core challenge during project quality initiation is to identify assumptions and risks. People from different work fields or different companies will approach a project with different unspoken assumptions. Moreover, many people tacitly assume that everyone interprets things the same way they do; this is often a faulty assumption that can lead to miscommunication and disappointment. Many risk events can be predicted. If risks are identified in advance and contingency plans are made, the effect on project quality is likely to be far less.

The fourth core challenge during project quality initiation is to identify lessons learned from previous projects that should be applied to the current project. While many organizations have come a long way in identifying lessons learned, they need to "close the loop" to benefit from them. A wise sponsor should not sign a project charter until the core team demonstrates that they have considered lessons learned from recently completed projects and found ways to incorporate that learning into the approach for the current project.

The final core challenge during project quality initiation is to have the project sponsor, project manager, and each member of the project core team personally commit to the project. This is normally accomplished through signing a project charter. The charter does not have to be detailed, but it does need to lay out some basic understandings and it does need to be considered a contract. When the sponsor, manager, and core team are all deeply and personally committed to accomplishing a project, it is more likely to be successful.

PROJECT QUALITY PLANNING CORE CHALLENGES

The five activities that pose core challenges during project quality planning are: determine customer satisfaction standards, determine customer tradeoff values, assess and prioritize improvement needs, have core team commit to the project plan, and have all key stakeholders commit to the project plan.

The first core challenge is to determine customer satisfaction standards. This activity is core because it is how the customers will judge whether the project deliverables are of acceptable quality. How can a project team expect to create good quality deliverables if they do not understand how the customer will judge the deliverables? This knowledge should form the basis of all the detailed planning that follows.

The second core challenge is to determine customer tradeoff values. To capture the customer's tradeoff values when the customer may not even fully know his or her values and may be unwilling to share them with the project manager requires some frank discussion. Unexpected events occur on most projects and understanding how the customer would make decisions regarding trading off a little of one project objective to gain more of another will enable the project manager and project team members to align their priorities with those of the customer.

The third core challenge during project quality planning is to develop the project quality management plan. This master plan links all prioritized process needs (including those that need to be created or improved) and resource requirements to the strategic priorities of both the project and the parent organization. This linkage enables all project participants to have clear goals, develop detailed responsibilities, and avoid suboptimization.

The fourth core challenge during project quality planning is for the core team to commit to the project plan. The core team members already committed to the project when they signed the charter. However, the charter was broad and the plan is much more detailed. Each member must be committed to the detailed approach that is now delineated.

The fifth core challenge during project quality planning is for all key project stakeholders to commit to the project plan. If the core team has not truly committed to the plan, this step will be very difficult. Key project stakeholders include internal and external direct purchasers, consumers, and providers. If any of these groups are not fully committed to the project, there is a greater chance of failure. Additionally, there are many other peripheral stakeholders who could disrupt the project. Ensuring that each of them understands the project and agrees with its goals (at least well enough not to attempt to interfere) is important.

PROJECT QUALITY ASSURANCE CORE CHALLENGES

The five core challenges during project quality assurance are: manage external customer quality assurance, manage internal customer quality assurance, perform ongoing review of process adequacy, conduct and report results of quality audits, and manage feedback changes.

The first core challenge is managing external customer quality assurance. This entails ongoing two-way communication between members of the project team and the customer. The purpose is to continuously convince all people in the customer organization that the project team members know what they are doing and will deliver results that will be useful to the customer.

The second core challenge is managing internal customer quality assurance. This includes ensuring that the project work is meaningful, workers are satisfied, and work systems are improving. On projects in which these activities occur, the organization gets stronger and its customers are better served.

The third core challenge in project quality assurance is conducting an ongoing review of project process adequacy. Figure 4-2, Project Quality Assurance Flowchart, shows eight in or out arrows connecting this activity with others. A great deal of information needs to be analyzed to ensure that work processes are still adequate. Many other downstream activities depend on the comprehensiveness and accuracy of that ongoing review.

The fourth core challenge in project quality assurance is conducting and reviewing the results of quality audits. The purposes of an audit are to ensure that everything is going according to plan and to suggest possible improvements. Audits can be viewed as a nuisance to be endured or as an asset to help make improvements. A prudent project manager takes the latter view.

The final core challenge in project quality assurance is to manage feedback changes. Projects encounter many changes for a variety of reasons. The important tasks to accomplish from a quality perspective are to identify possible changes, evaluate whether to implement them, and keep track of them. While most people understand the importance of those three tasks, when there is a great deal of time pressure, it is easy to lose discipline in this area. The key to accomplishing these three tasks is to make them simple so people will believe that the benefit of doing them outweighs the effort required.

PROJECT QUALITY CONTROL CORE CHALLENGES

The five core challenges during project quality control are: controlling project processes, approximating Six Sigma standards, using test results to correct deficiencies, having the project team endorse the deliverables, and having the customer accept the deliverables.

Controlling project processes is important because one of the key learnings from the TQM movement of the 1980s and 1990s is that quality cannot be inspected into products and services. Instead, quality must be designed and built in. Therefore, it is much more effective to control work processes so that the work is performed correctly the first time than it is to inspect afterward. Inspections do not find every defect and they take time and money—two commodities that are in short supply on most projects.

The second core challenge is to approximate Six Sigma standards. Many projects do not have enough repetition to document Six Sigma level quality, but the methods used to attain Six Sigma—identifying every opportunity for failure, assigning metrics to each, gathering data, analyzing root causes of problems, and gaining a firm commitment to produce as close to zero defects as possible—help enable a project team to produce as high quality a project as possible.

The third core challenge is to use test results to correct any defects. When things go wrong on projects, it is very tempting to guess the reason why and to "fix" it right away. There is often substantial time pressure. If a project uses good metrics, it should not take much longer to obtain the results of tests so the "fix" chosen is based on a combination of current data and the decision-maker's judgment rather than a rushed decision based solely on judgment. These databased decisions are more likely to be correct.

The fourth core challenge is for the project team to endorse the deliverables. If the project team members are to convince the customer that the deliverables are correct and complete, they need to believe it themselves. If there is a problem with any aspect of the deliverables, the team should discover it and correct it before trying to get the customer to concur.

The final core challenge in quality control is to have the customer formally accept the deliverables. This should be accomplished with truly satisfactory deliverables, and not under so much time pressure that the customer feels compelled to take imperfect deliverables. If the deliverables are complete and satisfactory, the customer can be an advocate for the project team in the future; if the deliverables are substandard, the customer can be a threat to the project team for future projects.

PROJECT QUALITY CLOSURE CORE CHALLENGES

Closure is considered an afterthought on too many projects. Often by the time a project is nearing completion, the participants have many other demands on their time and it is tempting to shortchange some of the project completion activities. Don't make this mistake! Core project quality closure challenges are: enabling customer capability; assessing the overall quality of all project processes; collecting, sharing, and documenting project lessons learned; recognizing and rewarding project participants; and obtaining referrals from capable and satisfied customers.

One of the most frequently overlooked activities during project quality closure is enabling customer capability. It is tempting to say "we know that the deliverables met the specifications and the customer was happy, so we are done." But you are not done until customers have had a chance to use the project deliverables under the entire range of operating conditions to verify their capability in using the deliverables without further training or support. Even if the customer has no money or desire to fund the training and support, the quality project organization will at least recommend to the customer the training and support they feel are needed.

Often the customer may need support and training for the useful life of the project deliverable. In these cases, the project of creating the deliverable needs to be declared complete at some point and a transition plan to the ongoing support and training needs to be developed and agreed to. What a project team does not want to happen is to complete the project, but have the customer feel it is poor quality because the training and support to fully use the project deliverable are not provided.

The second core challenge during project closure is assessing project work processes. Was each process effective (accomplishing what it was designed to), efficient (using no more time or resources than necessary), and adaptable (performing satisfactorily as project conditions changed)?

The third core challenge during project quality closure is to collect, share, and document all the lessons learned from the entire project. This can be simplified if the project team collects lessons learned at the end of each project stage. Collecting the lessons is only part of what is required. Categorizing the lessons, understanding how they all relate to each other, and ensuring that they are used in the future instead of merely sitting in a database somewhere are the other tasks that need to be accomplished to close the loop. Just as identifying lessons learned from previous projects is a challenge during project quality initiation, discovering how to ensure that future projects will benefit from the lessons learned on the current project is a challenge at project quality closure. The alternative is to keep relearning the same lessons.

The fourth core challenge is to recognize and reward project participants. Project participants need to be evaluated (preferably by several people who each worked with the participants on the project). Rewards and recognition should be given based on project performance. This includes both formal and informal recognition. First, there is a fairness issue of taking care of people who deserve it. Second, there is the motivational issue of people being willing to work harder if they feel they will be rewarded. A fair and caring project sponsor or project manager will have no trouble recruiting good, willing participants on future projects.

The final core challenge during project closure is obtaining referrals from satisfied, capable customers. One final measure of project quality is if the customer is willing to commit to future projects and to encourage others to do so as well. Only a satisfied and capable customer who is successfully using the project deliverables in his or her own business will be willing to do this. The reward is continuing, profitable business for the project organization.

Project Quality Participant Empowerment Readiness Assessment (PERA)

PURPOSE and DIRECTIONS

The purpose of the PERA is to measure the project quality empowerment readiness level of individuals and/or teams. It can be used as a 360 degree personal or team self-assessment instrument for prospective sponsors, project managers, and core team members.

For each of the dimensions listed below, circle the number that most closely represents your perception of the individual or team under consideration, using the rating scale below. Comments are optional.

	High	Moderate	Low
	8 7	6 5 4	3 2 1

Dimensions	Comments

1. Project technical credibility

Has technical project knowledge/credibility   Does not have project technical knowledge/credibility

8   7   6   5   4   3   2   1

2. Achievement motivation

Has high desire to achieve   Has low desire to achieve

8   7   6   5   4   3   2   1

3. Honesty

Is always honest   Is never honest

8   7   6   5   4   3   2   1

4. Quality problem-solving ability

Solves problems using quality tools   Is unable to solve problems using quality tools

8   7   6   5   4   3   2   1

5. Communication style

Communicates effectively at work				Does not communicate effectively at work				
8	7	6	5	4	3	2	1	_____

6. Trustworthy

Is always trustworthy				Is never trustworthy				
8	7	6	5	4	3	2	1	_____

7. Past project experience

Has relevant experience				Does not have relevant experience				
8	7	6	5	4	3	2	1	_____

8. Administrative knowledge/credibility

Always exhibits good operational judgment and tactful resourcefulness				Never exhibits good operational judgment and tactful resourcefulness				
8	7	6	5	4	3	2	1	_____

9. Justice/fairness

Is always fair and just				Is never fair and just				
8	7	6	5	4	3	2	1	_____

10. Quantitative knowledge/credibility

Uses quantitative skills effectively				Never uses quantitative skills effectively				
8	7	6	5	4	3	2	1	_____

11. Respectfully caring

Is always respectfully caring				Is never respectfully caring				
8	7	6	5	4	3	2	1	_____

12. Organizational conceptual knowledge

Knows the organization as a whole business system				Does not know the organization as a whole business system				
8	7	6	5	4	3	2	1	_____

13. Behavioral skills capability

Manages behavior of self and others at work effectively				Does not manage behavior of self and others at work effectively				
8	7	6	5	4	3	2	1	_____

14. Good moral judgment

Always exhibits good moral judgment				Never exhibits good moral judgment				_____
8	7	6	5	4	3	2	1	

15. Uses and shares power effectively

Uses and shares power effectively				Uses and shares power ineffectively				_____
8	7	6	5	4	3	2	1	

SCORING

Project Technical Task Maturity: Add the numbers circled for questions 1, 4, 7, 10, and 12, and divide the total by 5.

Project Administrative Psychosocial Maturity: Add the numbers circled for questions 2, 5, 8, 13, and 15, and divide the total by 5.

Project Participant Moral Maturity: Add the numbers circled for questions 3, 6, 9, 11, and 14, and divide the total by 5.

INTERPRETATION

Average scores for any of the factors:

0 - 4.0   = Individual or team is not ready for project quality empowerment at this time

4.1 - 7   = Individual or team is ready for regular participation in project quality teamwork

7.1 – 8.0 = Individual or team is ready for self-directed, high-performance project quality teamwork

Use your lowest average factor score as a place to begin preparing yourself or your team for responsible project quality empowerment. Individuals or teams who are prematurely empowered (e.g., individual selected to be a project manager without being ready to assume the commensurate responsibilities) eventually become problems for themselves, others, and the quality system (i.e., the "Peter Principle" of institutionalized incompetence).

Ethical Work Culture Assessment (EWCA)

PURPOSE and DIRECTIONS

The purpose of the EWCA is to determine the perceived level of moral development within the organization as a whole and the project team in particular. Think about what it takes for you and people like yourself (e.g., your co-workers, people in similar positions) to ì fit inî and meet expectations in your organization and in your particular project team. Select the number correlated with each response option below that best describes the current interpersonal behavioral styles of your organization and project team. Respond in terms of your perceptions of *how things are now,* not *how you would like them to be,* in both your organization and your project team. Place the number that correlates with each option in the appropriate blank spaces below under the columns labeled "organization" and "project team":

1 = Not at all	4 = To a great extent
2 = To a slight extent	5 = To a very great extent
3 = To a moderate extent	

SURVEY INSTRUMENT
Most people at work...

		Organization	Project Team
1.	Turn the job into a contest	_____	_____
2.	Play "politics" to gain influence	_____	_____
3.	Do things to avoid the disapproval of others	_____	_____
4.	Focus on pleasing those in positions of authority	_____	_____
5.	Involve others in decisions affecting them	_____	_____
6.	Trust that conflicts at work will be resolved fairly	_____	_____
7.	Appear hard, tough, and intimidating	_____	_____
8.	Oppose things indirectly	_____	_____
9.	Wait for others to act first	_____	_____
10.	Never challenge superiors	_____	_____

	Organization	Project Team
11. Resolve conflicts by majority vote	_____	_____
12. Help others to think for themselves	_____	_____
13. Compete rather than cooperate	_____	_____
14. Try to avoid appearing as a loser	_____	_____
15. Conform to the "way things are"	_____	_____
16. Treat rules as more important than ideas	_____	_____
17. Encourage and help others to participate in decision making at work	_____	_____
18. Demonstrate sincere caring for others at work	_____	_____
19. Maintain an image of superiority	_____	_____
20. Focus on building and maintaining a power base	_____	_____
21. Make "popular" rather than necessary decisions	_____	_____
22. Accord highest priority to respecting the "chain of command"	_____	_____
23. Think in terms of what would be supported by the majority of people	_____	_____
24. Try to "do the right thing" rather than "take the easy way "out"	_____	_____

SCORING

A. Add response numbers from questions 1, 7, 13, and 19   (Social Darwinism)

B. Add response numbers from questions 2, 8, 14, and 20   (Machiavellianism)

C. *House of Manipulation Score*   (total of A and B scores)   _____ _____

D. Add response numbers from questions 3, 9, 15, and 21   (Popular Conformity)

E. Add response numbers from questions 4, 10, 16, and 22   (Allegiance to Authority)

F. *House of Compliance Score* (total of D and E scores)   _____ _____

G. Add response numbers from questions 5, 11, 17, and 23   (Democratic Participation)

H. Add response numbers from questions 6, 12, 18, and 24   (Organizational Integrity)

I. *House of Integrity Score*
(total of G and H scores)   _____ _____

INTERPRETATION

Step 1: The highest total score among scoring steps C, F, and I indicates the level of moral development perceived by the respondent in both the organization and the project team. If C is the highest total score, the ethical work culture of the House of Manipulation predominates; if F is the highest score, the ethical work culture of the House of Compliance prevails; and if I is the highest total score, the ethical work culture of the House of Integrity prevails. Any level score ties are to be interpreted as indicating the lower (or lowest) work environment level of moral development.

The Organizational House is _____

The Project Team House is _____

Step 2: Once the work environment level has been determined (House of Manipulation, House of Compliance, or House of Integrity), the higher of the two scores that led to the level totals indicates the specific ethical work culture stage. Again, any stage score ties are to be interpreted as indicating the lower stage of moral development.

Work Environment Level Scores		Ethical Work Culture Stage	
If A is the higher total within the House of Manipulation	=	Social Darwinism	(Stage 1)
If B is the higher total within the House of Manipulation	=	Machiavellianism	(Stage 2)
If D is the higher total within the House of Compliance	=	Popular Conformity	(Stage 3)
If E is the higher total within the House of Compliance	=	Allegiance to Authority	(Stage 4)
If G is the higher total within the House of Integrity	=	Democratic Participation	(Stage 5)
If H is the higher total within the House of Integrity	=	Organizational Integrity	(Stage 6)

Organizational Ethical Stage is _____

Project Team Ethical Stage is _____

Step 3: Note any difference between organizational and project team scores since these disparities are points of both potential ethical conflict and opportunities for ethical work culture alignment and improvement. Persons caught between conflicting ethical work cultures for long periods of time experience severe work stress symptoms that inevitably impair optimal quality performance. Proactive management of ethical work culture development indicates human resource respect for people and commitment to building community at work.

Adapted with permission from Joseph A. Petrick and John F. Quinn, *Abbreviated Version of the Ethical Work Culture Assessment Instrument* (Cincinnati, OH: Organizational Ethics Associates, 1994).

Bibliography

Altman, Morris. *Worker Satisfaction and Economic Performance* (Armonk, NY: M.E. Sharpe, 2001).

Baker, Bud, and Raj Menon. "Politics and Project Performance: The Fourth Dimension of Project Management," *PM Network* 9, no. 11 (1995), 16–21.

Bechtold, Richard D. *Essentials of Software Project Management* (Vienna, VA: Management Concepts, 1999).

Belasen, Alan T. *Leading the Learning Organization* (Albany, NY: State University of New York Press, 2000).

Burnette, Donna K., and David Hutchens. "The New Face of the Project Team Member," *PM Network* 14, no. 11 (2000), 61-63.

CH2M HILL. *Project Delivery System* (Denver, CO: CH2M HILL, 1996).

Cleland, David I. *Project Management: Strategic Design and Implementation*, 2nd ed. (New York: McGraw-Hill, 1994).

Cleland, David I. (Ed.) *Field Guide to Project Management* (New York: Van Nostrand Reinhold, 1998).

Cleland, David I., James M. Gallagher, and Ronald S. Whitehead. *Military Project Management Handbook* (New York: McGraw-Hill, 1993).

Cleland, David I., and Lewis R. Ireland. *Project Manager's Portable Handbook* (New York: McGraw-Hill, 2000).

Cochran, Dick. "Finally, a Way to Completely Measure Project Manager Performance," *PM Network* 14, no. 9 (2000), 75-80.

Corbin, Darrell, Roger Cox, Russ Hamerly, and Kenneth Knight. "Project Management of Project Reviews," *PM Network* 15, no. 3 (2000), 59-62.

Crosby, Philip B. *Quality Is Free: The Art of Making Quality Certain* (New York: Dutton, 1979).

Darnall, Russell W. *Achieving TQM on Projects: The Journey of Continuous Improvement* (Upper Darby, PA: Project Management Institute, 1994).

Deming, W. Edwards. *The New Economics for Industry, Government, and Education* (Cambridge, MA: MIT Center for Advanced Engineering, 1993).

Deming, W. Edwards. *Out of the Crisis* (Cambridge, MA: MIT Center for Advanced Engineering, 1986).

Englund, Randall L. "Capturing Project Requirements and Knowledge," *PM Network* 14, no. 2 (2000), 49-59.

Eisner, Howard. *Essentials of Project and Systems Engineering Management* (New York: John Wiley & Sons, 1997).

Evans, James R., and James W. Dean, Jr. *Total Quality: Management, Organization, and Strategy*, 2nd ed. (Cincinnati, OH: South-Western Publishing, 2000).

Evans, James R., and William M. Lindsay. *The Management and Control of Quality*, 5th ed. (Cincinnati, OH: South-Western Publishing, 2002).

Githens, Gregory D. "Capturing Project Requirements and Knowledge," *PM Network* 14, no. 2 (2000), 49-59.

Gryna, Frank M. *Quality Planning and Analysis: From Product Development through Use* (Boston: McGraw-Hill Irwin, 2000).

Harrington, James S. *Total Improvement Management: The Next Generation in Performance Improvement* (New York: McGraw-Hill, 1995).

International Organization for Standardization (ISO). *Quality Management and Quality Assurance* (Geneva, Switzerland: ISO Press, 1994).

Ireland, Lewis R. *Quality Management for Projects and Programs* (Upper Darby, PA: Project Management Institute, 1991).

Jacobson, Ivar, Grady Booch, and James Rumbaugh. *The Unified Software Development Process* (Reading, MA: Addison Wesley, 1999).

Juran, J. M. *Juran on Leadership for Quality* (New York: Free Press, 1989).

Juran, J. M. (Ed.) *A History of Managing for Quality* (Milwaukee, WI: ASQ Press, 1995).

Kerzner, Harold. *In Search of Excellence in Project Management* (Glastonbury, CT: International Thomson Publishing Company, 1998).

Kerzner, Harold. *Project Management: A Systems Approach to Planning, Scheduling, and Controlling*, 7th ed. (New York: John Wiley & Sons, 2001).

Kirkpatrick, D.L. *A Practical Guide for Supervisory Training and Development* (Reading, MA: Addison-Wesley, 1971).

Kloppenborg, Timothy J. "Project Management," *Encyclopedia of Business*, 2nd ed. (Detroit, MI: Gale Research, 1999), 803-806.

Kloppenborg, Timothy J., and Samuel J. Mantel. "Project Management," *The Concise International Encyclopedia of Business and Management*, 2nd ed. (London: Thompson Press, 2001), 560-565.

Kloppenborg, Timothy J., and Samuel J. Mantel. "Tradeoffs on Projects: They May Not Be What You Think," *Project Management Journal* 32, no. 1 (1990), 38-53.

Kloppenborg, Timothy J., and Joseph A. Petrick. "Leadership in Project Life Cycle and Team Character Development," *Project Management Journal* 30, no. 2 (1999), 8-13.

Kloppenborg, Timothy J., and Joseph A. Petrick. "Meeting Management and Group Character Development," *Journal of Managerial Issues* 11, no. 2 (1999), 166-179.

Kloppenborg, Timothy J., Warren A. Opfer, Peter Bycio, Julie Cagle, Thomas Clark, Margaret Cunningham, Miriam Finch, James M. Gallagher, Joseph Petrick, Rachana Sampat, Manar Shami, John Surdick, Raghu Tadepalli, and Deborah Tesch. "Forty Years of Project Management Research: Trends, Interpretations, and Predictions," *Proceedings of PMI Research Conference 2000: Project Management Research at the Turn of the Millennium* (Project Management Institute, 2000), 41-59.

Knarbanda, O.P., and Jeffrey K. Pinto. *What Made Gertie Gallop?: Learning from Project Failures* (New York, NY: Van Nostrand Reinhold, 1996).

Knutson, Joan. "Developing a Team Charter," *PM Network* 11, no. 8 (1997), 15-16.

Knutson, Joan. "That First Step Can Be the Most Important," *PM Network* 13, no. 9 (1999), 19-20.

Knutson, Joan. "Will the Real Project Client Please Stand Up?" *PM Network* 15, no. 4 (2001), 26-27.

Leavitt, Jeffrey S. and Philip C. Nunn. *Total Quality through Project Management* (New York: McGraw-Hill, 1994).

Lee, Thomas H., Shoji Shiba, and Chapman Wood. *Integrated Management Systems: A Practical Approach to Transforming Organizations* (New York: John Wiley & Sons, 1999).

Lindsay, William M., and Joseph A. Petrick. *Total Quality and Organization Development* (Delray Beach, FL: St. Lucie Press, 1997).

MacMaster, Gornon. "Can We Learn from Project Histories?" *PM Network* 14, no. 7 (2000), 66-67.

Melan, Eugene H. *Process Management: Methods for Improving Products and Services* (New York: McGraw-Hill, 1992).

Meadows, Dennis. "The TQM Vital Signs of a Project," *1998 Proceedings of the Project Management Institute* (1998), 18-20.

Meredith, Jack R., and Samuel J. Mantel, Jr. *Project Management: A Managerial Approach*, 4th ed. (New York: John Wiley & Sons, 2000).

Mickelson, P., and S. Elliot. *Construction Quality Program Handbook* (Milwaukee, WI: American Society for Quality Control, 1986).

Nicholas, John M. *Competitive Manufacturing Management* (Chicago, IL: McGraw-Hill, 1998).

Nonaka, Ikujiro, and Hirotaka Takeuchi. *The Knowledge-Creating Company* (Oxford: Oxford University Press, 1995).

Oswald, Thomas H., and James L. Burati. *Guidelines for Implementing Total Quality Management in the Engineering and Construction Industry* (Clemson, SC: Clemson University Press, 1992).

Petrick, Joseph A., and Diana S. Furr. *Total Quality in Managing Human Resources* (Delray Beach, FL: St. Lucie Press, 1995).

Petrick, Joseph A., and John F. Quinn. "The Challenge of Leadership Accountability for Integrity Capacity as a Strategic Asset," *Journal of Business Ethics* 34 (2001), 331-343.

Petrick, Joseph A., and John F. Quinn. "The Integrity Capacity Construct and Moral Progress in Business," *Journal of Business Ethics* 23 (2000), 3-18.

Petrick, Joseph A., and John F. Quinn. *Management Ethics: Integrity at Work* (Thousand Oaks, CA: Sage, 1997).

Pinto, Jeffrey K. *Power and Politics in Project Management* (Upper Darby, PA: Project Management Institute, 1996).

Project Management Institute Standards Committee. *A Guide to the Project Management Body of Knowledge (PMBOK® Guide)* (Upper Darby, PA: Project Management Institute, 2000).

Rizzo, Tony. "Operational Measurements for Product Development Organizations," *PM Network* 13, no.11 (1999), 42-47.

Scholtes, Peter, R., Brian L. Joiner, and Barbara J. Streibel. *The Team Handbook*, 2nd ed. (Madison, WI: Joiner Associates, 1996).

Shiba, Shoji, Alan Graham, and David Walden. *A New American TQM: Four Practical Revolutions in Management* (Portland, OR: Productivity Press, 1993).

Stevens, James D., Timothy J. Kloppenborg, and Charles R. Glagola. *Quality Performance Measurements of the EPC Process: The Blueprint* (Frankfort, KY: University of Kentucky, 1994).

Stewart, Thomas A. *Intellectual Capital: The New Wealth of Organizations* (New York: Currency, 1997).

Stewart, Wendy E. "Balanced Scorecard for Projects," *Project Management Journal* 32, no. 1 (2001), 38-53.

Sveiby, Karl E. *The New Organizational Wealth: Managing and Measuring Knowledge-based Assets* (San Francisco, CA: Berrett-Koehler, 1997).

Swanson, Roger C. *The Quality Improvement Handbook* (Delray Beach, FL: St. Lucie Press, 1995).

Swift, J.A. (1995). *Introduction to Modern Statistical Quality Control and Management* (Delray Beach, FL: St. Lucie Press, 1995).

Turner, J. R. *The Handbook of Project-based Management* (London: McGraw-Hill, 1992).

VanEpps, David E. "Setting Expectations: Initiating the Project Manager/Client Relationship," *PM Network* 14, no. 9 (2000), 101-102.

Verma, Vijay K. *Organizing Projects for Success: The Human Aspects of Project Management* (Upper Darby, PA: Project Management Institute, 1995).

Wheeler, Donald J. *Understanding Variation: The Key to Managing Chaos* (Knoxville, TN: SPC Press, 1993).

Zairi, M. *TQM-Based Performance Measurement: Practical Guidelines* (London: Technical Communications Ltd., 1992).

Zeitoun, Alda A., and Garold D. Oberlender. *Early Warning Signs of Project*

Index

A
abnormal variation, 54
acceptance sampling techniques, 2
agenda, project kick-off meeting, 58
American National Standards Institute (ANSI), 2
American Society for Quality (ASQ), 2
assessments, 91
assignable variation, 15
assumptions, identifying, 26, 32
assurance. *See* project quality assurance
attributes, 79
audits, internal and external, 31

B
Bell Telephone Laboratories, 1–2
benchmarking, 50, 81
breakthrough dominance, 13
broad targets, 28–30
budget, 8, 49

C
cause-and-effect diagram, 48–49
closure. *See* project quality closure
co-partnership, 90–91
combination, 35
commitment
    to plan, 44, 56
    to project, 26, 38–40
common variation, 15
company-wide quality control (CWQC), 2
competence, 17
competitive parity, 13
concurrent engineering, 51
control. *See* project quality control
control charts, 2, 80
control limits, 79
core challenges
    project quality assurance, 99–100
    project quality closure, 101–102
    project quality control, 100–101
    project quality initiation, 97–98
    project quality planning, 98–99

core project team
    kick-off meeting, 58
    project communications plan, 54–55
    selecting, 26, 30
cost/benefit analysis, 50
cost of poor quality, 6–7
cost/quality tradeoffs, 29
cross-functional teams, 16
CSS. *See* customer significance-success matrix
customer capability, 90–91
customer satisfaction
    importance of, 8, 10–12, 19
    project quality assurance, 64–66
    project quality closure, 87, 90–91
    project quality control, 78–80
    project quality initiation, 26–31
    project quality planning, 41, 44–47
customer significance-success (CSS) matrix, 65–66
customer standards matrix, 44–45
customer tradeoffs, 45–47, 90
customers, 11
CWQC. *See* company-wide quality control

D
data and measurement matrix, 54–55
decision-making authority, 44, 47
delays, 68
deliverables, 78, 85
Deming, W. Edwards, 2
Deming Prize, 2
Dilbert effect, 16
dissatisfiers, 12

E
economic analysis tools, 2
efficiency problems, 82
EFQM. *See* European Foundation for Quality
    Management
empowered performance
    importance of, 17–18, 20
    project quality assurance, 64, 71–72
    project quality closure, 87, 93